Your Brain and You

A Simple Guide to Neuropsychology

Dr Nicky Hayes, C. Psychol. FBPsS

Contents

Preface

This book is all about how our brains work to make us who we are. As research into the brain continues, we are discovering more and more about what parts of the brain do what, and how all those various parts connect together. It's not simple: what we do, think, say and feel comes from many different brain areas all working together. Somehow, these areas work in concert to make us who we are. But that doesn't mean that we are only a collection of nerve impulses. Our brains affect what we do, but what we do can change our brains, as well. Who we are, as people, is to do with the choices and decisions that we make in our lives. What we are exploring here is how our nerve cells and brain structures make those choices possible.

Another aspect of modern brain science is that it teaches us about neurodiversity, that is, that everyone is different. Each brain is slightly different, we all have different talents and skills, and sometimes a set of characteristics that we think of as 'normal' might actually be quite uncommon. You'll read about how one area of the brain can be doing something while another area simultaneously does something else. But it's important to remember that such simple descriptions are never the full story: brains can change; established pathways can be interrupted and regrow using different cells; some people are born with brains that do unusual things; and none of us is identical. But we are all marvellous! Our brains are amazing, and I hope this book will help to show you just how amazing they are.

Nicky Hayes

What is the brain?

In this chapter you will learn:

- ▶ *how the human brain evolved*
- ▶ *the elements of the basic brain*
- ▶ *how the thalamus and the limbic system work*
- ▶ *the function of the cerebrum.*

What makes human beings special? People have put forward any number of answers to that question. They have suggested that it is our ability to tell stories, to work together, to store information, to laugh, to imagine, to use language, to learn or to solve problems. It has even been suggested that we are distinctive because we are not really distinctive: we don't have specialized horns, teeth, other natural weapons or the ability to run fast, and although we can do a lot of physical things, there are almost always other animals that can do them better. So because we are not particularly specialized in terms of our physical abilities or attributes, we have to work out different ways of doing what we need to do.

Any or all of these may be true. But underlying them all is the one thing which makes it all possible: the very special brain that human beings have evolved, and the way that it allows us to interact with our worlds – our physical worlds, our social worlds and our imaginary worlds. That brain is something very special, and it is what allows human beings to be what we are.

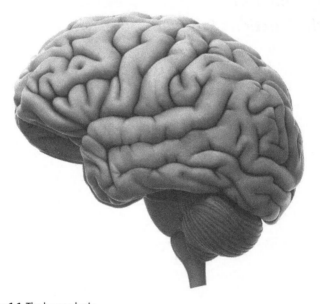

Figure 1.1 The human brain

Our brains allow us to see things and make sense of what we are doing. They allow us to take action: to move purposefully and do things when we need to or want to. They allow us to hear: to interpret vibrations in the air, to identify where they are coming from, and to identify their likely cause. They do the same for all of our other senses, including the sense receptors we have inside our bodies, which tell us what our muscles and joints are doing. Our brains make it possible for us to locate ourselves in the material world: to receive information from it, and to act within it.

But they do much more than that. Our brains also allow us to remember things – and in more than one way. They store conscious memories, like PIN numbers and addresses, but they also allow us to remember things that happened in the past, and even allow us to remember to do things in the future (most of the time!). They allow us to store skills, so that we can perform actions or cognitions smoothly and without consciously thinking about the steps involved; and they store patterns and meanings, so that we can make sense of new things that we encounter. They even allow us to imagine things that might happen in the future – or things that might never happen.

As social animals, it is important that we are able to recognize people, and it is our brains that allow us to recognize faces and bodies, and to distinguish between familiar and unfamiliar individuals. Our brains also make it possible for us to develop the attachments and relationships that are the basics of social living and to communicate with other people, using words, signs or symbols. At a more abstract level, our brains also make it possible for us to deal with the 'three Rs' – reading, writing and arithmetic, each of which involves distinct areas of the brain. But being human is more than just having mental skills of this kind: it is our ability to empathize with others which really makes us human, and our brains also provide us with the mechanisms for self-knowledge, identification and empathy.

We have emotions, too, and these are only possible because of how our brains have evolved. We feel anger, fear, happiness and disgust, we feel pleasure and pain, and we respond to rewards. We have times when we are alert and agitated, times when we are relaxed or experiencing states like mindfulness, and times when we are asleep. These states of consciousness are part of how our brains work. And, as human beings living modern lives, we also make decisions. The human brain is able to cope with decisions at various levels, ranging from deciding to sip a coffee to deciding to buy a house. The brain is an amazing structure, and in this book we will be exploring all these aspects of how it works.

How did the brain begin?

How did our brains become so complex? Back in evolutionary history, the first animals didn't have brains at all. They were simple, one-celled organisms, a bit like the amoebae of today, which float in their liquid environment and absorb particles of food as they come across them. As more complex animals developed, one of their advantages was that they became able to detect nearby sources of food. They began to develop specialized cells that could identify the chemical changes in their environment produced by nearby food; and other cells that would help to propel their bodies towards it. They also developed a central linking system, which allowed them to use the information they were receiving and direct their movement accordingly. That central linking system acted as a co-ordinator between the incoming information and the resulting action.

And that was the beginning of it all. The first nervous systems were a simple, ladder-like network of fibres through the body, linked to a simple tube – which we call the **neural tube**. There's something similar in the bodies of modern-day planaria, or flatworms. It's basic, but we know it works because they still survive today. As animals became more complex, so did the structure of the nervous system. The front end of the neural tube began to become enlarged: it was the co-ordination centre which received information from the detectors that identified possible food, or light, or other information like vibrations, which implied that something large was nearby. Those detectors eventually became the sense organs, and the enlarged front part of the neural tube became the brain. The rest of the tube, which passed along the body, became the spinal cord, and the cells that passed information to and from it became the somatic (bodily) nerves. But even though it became so much more elaborate, it was, and still is, a kind of tube. It just has many more knobbly bits on the end than a flatworm has.

By the time of the dinosaurs, animals had become much more complex. That swelling at the front end of the neural tube had become a brain – not a very big one but one with different parts, which allowed it to co-ordinate the various bodily mechanisms needed to keep the animal alive, such as respiration, digestion and heartbeat. The brain also received information from the senses, which had become much more sophisticated, with their own separate organs and nerves and their own specialized parts of the brain. Movement and balance, too, had become vitally important functions, and a large part of the brain had developed to deal with these. And even a type of memory, although less complex than the memory we use, had begun to evolve. The dinosaur brain was tiny by comparison with our modern human brain, but, as palaeontology shows, it worked, and it worked very well.

Dinosaurs dominated the land for many millions of years, and their descendants, the birds, are still with us today.

The same brain developments applied to other animals, like fish, amphibians and reptiles. Their adaptation to different ecosystems and food sources led them to evolve into all sorts of different creatures. Some of those ecosystems encouraged them to develop a highly sophisticated sense of smell, so the part of the brain dealing with smell became enlarged. Others required acute vision, which meant that the part of the brain dealing with vision became enlarged. Some animals developed an acute sensitivity to vibrations in the air, leading to an enlarged brain centre for hearing, and so on. As animals evolved to deal with their environments, so the brain evolved to co-ordinate that adaptation.

Another set of animals had appeared during the time of the dinosaurs: the mammals. These had evolved another special part of their brains, which was able to control and regulate their body temperature. As a result, mammals could be active at night and avoid the reptile predators that depended on the sun for warmth and energy. Mammals evolved in other ways, too: they began to suckle their infants and nurture them after they were born, which allowed the young animals a safe time to learn about the physical world around them and to explore. A small part of the mammals' brain became specialized for adaptation and learning, so they were able to deal with unpredictable or changing environments. All this meant that when the world changed and the dinosaurs died out, the mammals were able to survive and take advantage of the ecological resources the dinosaurs were no longer using.

The mammalian brain, like that of other animals, adapted itself to the demands of its environment. Prey animals became highly sensitive to sensory information, developing acute reflexes enabling them to react quickly. Hunting animals developed in similar ways, as their survival required them to match the prey animals in order to catch them. Some animals were vegetarian, living only on plants; others were omnivores, exploiting whatever food sources they could find. And, most important of all, some of these lived socially, and shared their resources.

Since living socially was important, the demands of social interaction and co-operation meant that mammals lived in an ever-changing environment, so those parts of the brain that allowed them to deal with change and to communicate and pass on information became well developed. Animals living in social groups therefore developed multipurpose brains, which could adapt to different environments, interact with different individuals, spot new opportunities, and deal with problems. And in one particular group of mammals it became so important that it eventually overshadowed all the other parts.

When we look at a human brain today, almost all we can see are the two halves of the cerebrum: the part of the brain which we use for thinking, learning, communicating, deciding, imagining, and just about everything else which makes us human. The other, older parts of the brain are still there, but the cerebrum is so large that it has expanded to spread all over the rest. And since it's the outer skin, or cortex, of the cerebrum which does most of the work, the cerebrum has become folded and wrinkled, so we can fit more surface area into the space. The human brain is one of the most remarkable things we know about, and understanding fully how it works is likely to keep our scientists busy for many generations to come.

The basic brain

In this book we'll be exploring what modern neuroscientists have been able to find out about the ways that the brain functions. But even before we do that, we can learn a lot just from looking at the different parts of the brain and seeing how each part has evolved. Let's begin by thinking about the most basic nerve functions a developing animal would need: to be able to move and to avoid pain.

This takes us back to the ancient neural tube. We still have the equivalent of it in our human central nervous system, although it's become a bit more sophisticated since then, of course. It's the **spinal cord** – the tube of nerve fibres running the length of the spine, linking our body's nerve fibres with the brain. If we look at a cross-section of the spinal cord, we can see that it really is a tube: it has a hollow in the middle that is filled with a nutrient fluid. That hollow is surrounded by what we call grey matter, which is mostly made up of nerve cell bodies, and the grey matter is surrounded by white matter, which is nerve fibres carrying information to and from the brain. So the spinal cord is the main route for information going between the body and the brain. That's why people who experience damage to their spinal cord can become paralysed. Their brains may be trying to get their muscles to move, but they simply cannot get the instructions through.

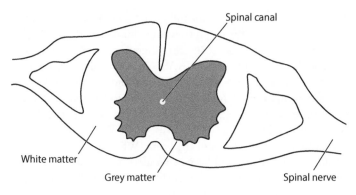

Figure 1.2 The spinal cord in cross-section

Not all movement is directed by the brain, though. The spinal cord also controls some of our **reflexes** – the rapid muscle movements that happen in response to painful stimuli. This is what happens, for example, if you pull your hand away from a hot surface. You do it quickly, without thinking, because the message of heat and pain only needs to go as far as the spinal cord. When it gets there the message is instantly routed to the nerve cells, which tell your arm to move your hand away. It doesn't need to go all the way up to the brain. That's known as a reflex and, because it's such a basic survival mechanism, it's controlled by the oldest part of the nervous system. What happens is that the message – pain, unexpected pressure, or whatever – is picked up by sensory nerve cells, which pass the information up to the nerve cells in the spinal cord. From there, a message is immediately passed on to motor nerve cells, which connect to the muscles and instruct them to contract. So you pull your hand away, or jerk your foot upwards, or respond in whichever way the reflex is appropriate.

The top of the spinal cord thickens out and begins to become part of the brain itself. The part where it thickens is known as the **medulla,** and if we think about it as the next part of the nervous system to evolve, we can see how it, too, is concerned with basic functioning. The medulla is the part of the brain that regulates basic bodily functions, such as breathing, swallowing, digestion and heartbeat – essential functions for all animals apart from the very simplest ones.

The basic brain is therefore a system that allows an animal to move, react to pain, eat, breathe and pass nutrients around its body. But if that animal is to survive in an increasingly complex world, it also needs to be alert and ready to move if something threatens it. Moving upwards from the medulla, we find that the brainstem becomes even thicker, turning into what is known as the **midbrain**, which is really a collection of several different parts. One of them is the reticular activating system (RAS), which regulates different states of alertness: sleep, wakefulness and attention. In humans and complex mammals, the RAS seems to be able to 'switch on' large areas of the cerebral cortex, so we are alert and paying attention to what is around us. It has some sensory pathways and many connections with other areas of the brain. When we think about its evolutionary origins, and how being alert would have helped an animal to survive in a dangerous world, we can see why this was quite an early part of the brain to evolve.

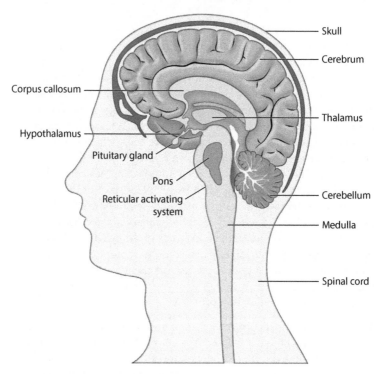

Figure 1.3 The structures of the brain

Other areas of the midbrain include the superior colliculi and the inferior colliculi (superior means above and inferior means below, which tells you how they are placed). These are oval structures which do some very basic sensory processing: the superior colliculi are particularly concerned with vision and touch, while the inferior colliculi are mainly specialized for hearing. They don't connect directly with the higher levels of the brain. Instead, they have a direct connection with our attention and movement systems, alerting us immediately if there is, say, a sudden flash or bang. You can see how that would also have helped an animal to survive.

As the brain evolved, animals were also becoming more sophisticated in how they moved. Another part of the midbrain, the pons, is the main route for connections between the body and the cerebellum, which is the co-ordinating centre for smooth movement. The pons is also involved in dreaming sleep, in animals as well as humans, and it is thought that this might have evolved to help the animal to form the neural pathways needed for smooth movement. When we see dogs dreaming, for example, it's evident that they are running or chasing something, which could possibly link with practising physical skills. We'll come back to how mental rehearsal can help skill learning in Chapter 6.

The pons is connected to the **cerebellum**, which is the wrinkled bulge that sticks out underneath the back of the cerebrum. It's sometimes called the mini-brain, and it is capable of carrying out many more complex functions than just keeping the animal alive. As we've seen, it plays an important role in skill learning. When we are first learning a new skill, our movements are often jerky and a bit clumsy, because we have to think about each movement consciously. But as we practise those movements, control of those sequences of actions moves to the cerebellum, and those movements become smooth and automatic, so we don't have to think about doing them. The cerebellum doesn't plan deliberate movement – that's done by the cerebrum – but it makes sure that our actions are co-ordinated, precise and accurately timed.

Like the cerebrum, the cerebellum is organized into two halves, and its surface is folded closely like the leaves of an accordion. This suggests that the outer layers are particularly important for the cerebellum's functioning (the grooves and folds increase the total surface area). The folds mean that its surface comprises most of its structure, but there are nerve fibres underneath them and a small space filled with fluid, known as a ventricle, at the point where it joins the pons. The cerebellum also controls balance, which is a function that appears to be located in a small knob between its two halves. Most drugs aiming to treat motion sickness have the effect of suppressing this area of the brain. In humans, the cerebellum is also included in some of the nerve pathways involved in processing attention, language and fear and pleasure reactions. So you can see that it is an important part of the brain for all complex animals.

These structures, collectively, support essential bodily processes, so we can understand why they would have evolved first.

The thalamus and the limbic system

Later on, even more complex structures began to evolve in the brain. While many simpler animals had sensory organs of some kind, reacting to light or vibrations or changes in the chemical composition of the fluid they were living in, some began to develop more sophisticated perception and evolved brain structures to deal with it. For example, there is a large area of packed cells above the midbrain and below the cerebrum known as the **thalamus**. It is separated into two halves, and it acts as a kind of relay station for sensory information, and for motor signals going to the muscles.

The thalamus receives information from the sensory nerves and from our eyes and ears, and does a certain amount of decoding of those signals before passing the information on to the cerebrum. It also receives the instructions about movement passed down from the cerebrum and sends those instructions on to our muscles. Like several other subcortical structures, it's involved in sleep and wakefulness as well – those states seem to affect large areas of the brain in a general way rather than being tightly controlled by just one area.

There are a number of other small structures around the thalamus, collectively known as the **limbic system**. A small 'lump' immediately below the thalamus, known as the **hypothalamus**, is especially important to mammals because it regulates body temperature. It's the ability to keep our own internal body temperature constant that allows us to be active at night or in cold places. It is also the reason why small mammals can live in underground burrows, which is a possible explanation for how they were able to survive the massive impact cataclysm that finished off the dinosaurs.

The hypothalamus does much more than just regulate temperature, though; it maintains **homeostasis** throughout the body. Maintaining homeostasis means keeping everything in a steady, comfortable condition. So if your body's fluid levels fall below what is optimal for your survival, the hypothalamus will initiate feelings of thirst leading you to drink; if your body's blood glucose levels fall below a certain level, it will initiate hunger, leading you to seek food. If you get too cold, it initiates shivering, which agitates your muscles to generate a little heat; and if you get too warm, it initiates

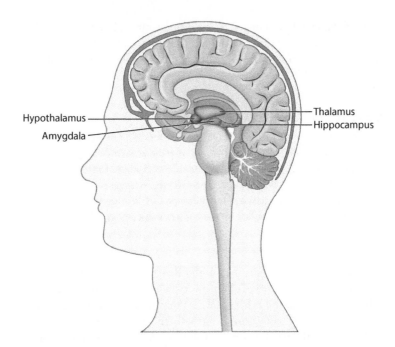

Figure 1.4 The elements of the limbic system

sweating, so that the evaporation will cool you down. Effectively, the hypothalamus acts as the regulator to keep the body's basic mechanisms working as they should. Its position just below the thalamus ('hypo' means 'below') allows it to have connections with all the early parts of the brain, so it can send the relevant signals when they are needed.

The hypothalamus sends its signals partly by nerve cell (neural) connections, but partly also by releasing **hormones**. Hormones are chemicals that either stimulate body processes or cause other hormones to be released by other glands in the body. Hormones are particularly important for maintaining 'states' such as growth, pregnancy, arousal or anxiety. Together, the hormone-releasing glands form the endocrine system of the body, and the hypothalamus is the brain's main route for connecting the brain with the endocrine system.

There are many other parts of the limbic system. The hippocampus is a small curved structure underneath the cerebrum, and its name comes from the way that it was thought to resemble the shape of a seahorse. It's important in memory in several ways. One is that it allows us to consolidate our memories into long-term memory storage. People with damage to the hippocampus, like that caused by frequently drinking alcohol without eating, can find that they become unable to store new memories. This is known as Korsakoff's syndrome, and it can have tragic consequences.

Case study: Korsakoff's syndrome

A case reported by the neurologist Oliver Sacks was of a 60-year-old man who had developed Korsakoff's syndrome when he was about 30 years old. The damage to his hippocampus meant that he was unable to store any new memories: he retained only the memories from his younger self. This meant that every day when he looked in the mirror he had an unpleasant shock, not recognizing the old man he had become or remembering anything of his more recent past.

The **hippocampus** is concerned with other forms of memory as well – for example our spatial memory, which is how we remember where we are, and the other locations that we know. London taxi drivers, who have to memorize virtually the whole of London in order to pass the test known as 'The Knowledge', have shown an increase in the size of the hippocampus as a result of their increased spatial memory. So what we do in life can either damage how our brain works or improve it. It's all about the choices we make. We'll look at how memory works in the brain in more detail in Chapter 7.

From the evolutionary perspective, though, we can see how the ability to navigate round an area and to develop mental maps can help an animal to survive. Studies have shown, for instance, that mice allowed to explore a complex maze with no escape routes will freeze if a cat is then introduced to the area, while mice that haven't had a chance to explore will run about looking for an escape. By exploring and remembering, the first set of mice discovers that escape isn't possible, so staying very still to avoid attention is a better option. Incidentally, we don't know whether the cat was allowed to catch the mice in these studies, or what happened to them in the end. In the period when they were conducted – the 1960s – little consideration was given to animal ethics, so the mice might or might not have survived. We wouldn't do a study like that nowadays, fortunately! But it doesn't change the main point, which is that knowing all about your location is a definite aid to survival.

Another important part of the limbic system is the **amygdala**. This is the emotion centre of the brain, and it consists of two almond-shaped structures located deep in the right and left temporal lobes, quite close to the hippocampus. It helps the brain to identify and react to threats, and is active in our other emotions, too – both positive and negative ones. Part of its role appears to be working with the hippocampus to consolidate memories, especially emotional ones. We are more prepared to remember things if they accompany an intense emotional experience, and this is partly the result of activity in the amygdala. We'll come back to the actions of the amygdala in many of the chapters in this book, but particularly in Chapter 8.

Other parts of the limbic system include the **basal ganglia,** which is the name given to a group of cells nested deep in the white matter of the frontal lobes. These cells help us organize our movements by choosing appropriate actions, and inhibiting our actions until we know that they are suitable for the situation. The basal ganglia also include the caudate nuclei, which are also involved in planning actions and in co-ordinating the learning of habits and rule-based actions, and the area known as the globus pallidus, which is all about regulating deliberate movement so that it is co-ordinated and fluid. As you might expect, then, the basal ganglia have close links with the cerebellum, and damage to any of these areas can produce problems with movement of one kind or another.

The cingulate gyrus, or **cingulate cortex**, is a large area of the brain just above the corpus callosum. Although it is continuous with the cerebrum itself, it is often considered to belong to the limbic system, partly because of its connections and the way it works so closely with other parts of that system, like the hypothalamus and the amygdala. It is involved in emotions, memory and learning: among other functions, for example, it co-ordinates smells and sights with pleasant or unpleasant memories. It also seems to be involved in regulating aggressive behaviour, and it is active in the neural pathways that are stimulated by our emotional reactions to pain.

The limbic system, then, is intensely involved in emotions, memory and movement – all of which are important for animals like mammals, which need to be able to survive in a complex world. Other animals – reptiles, fish and amphibians – also possess these structures but we are not as clear about the precise roles they play. Our understanding of mammals, in terms of emotions and their learning mechanisms, is better developed, partly because it helps us to understand how our own nervous system has developed.

Remember this: Folds and grooves

Information is processed on the outer layer of the cerebrum – the cerebral cortex. In some animals, like birds or reptiles, this outer surface is relatively smooth and the cerebrum itself is not particularly large. In mammals the cerebrum is bigger, with creases and grooves, which increase its surface area. The cerebrum is the largest part of a dog or cat's brain, and it has several deep folds. A monkey's cerebrum is much larger than a cat's, relative to the rest of its brain, and has many more folds and grooves. Apes have even more. By the time we reach humans, the cerebrum covers up almost all of the rest of the brain and its surface has become very convoluted indeed, with parts which can't even be seen from the outside because they are folded right underneath. This fits neatly with our ideas about humans being more intelligent than other animals – until, that is, we look at whales and dolphins, which have even more folds and grooves in the surface of their cerebrum than humans do. Does that mean they are more intelligent than us? Nobody knows, except possibly the cetaceans, and they aren't telling.

The cerebrum

Finally, moving upwards through the brain, or onwards through our evolutionary progression, we come to the **cerebrum**. This is the largest structure of all in mammals, and especially in humans (and cetaceans too, although we don't really know what whales and dolphins do with their massive brains!). It is by far the most important part of the brain in human beings, and most of this book will be dealing with different aspects of our cerebral functioning. For now, though, it is worth taking a look at its basic structure, so that we can find our way around its different areas and sections.

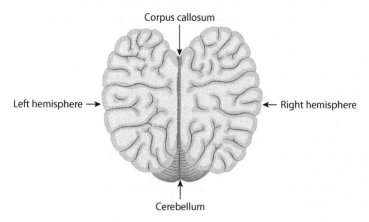

Corpus callosum

Left hemisphere →   ← Right hemisphere

Cerebellum

Figure 1.5 The cerebral hemispheres seen from above

The cerebrum is what makes us human. It is the part of the brain that does thinking, perceiving, language, imagining and planning, decision-making, socializing, and all the other aspects of cognitive and social understanding that we use without even being aware of it. So it's not surprising that it overshadows all the other structures. It isn't separate from them: the surface of the cerebrum is composed of grey matter – or gray matter, if you want to spell it that way – and it consists of the cell bodies of neurones and the other cells which support them. But underneath the layer of grey matter is a compacted mass of white matter, which consists of the long fibres, or axons, connecting those neurones with other parts of the cerebrum and other parts of the brain. Our brains are full of pathways of nerve fibres, linking all the different parts together, and this book is about what neuroscientists have been able to discover about the ways these pathways work.

The cerebrum itself is divided into two halves, a bit like a giant walnut. The two halves are the left and right **cerebral hemispheres**. They are largely separate, but they have a crossover, a thick band known as the corpus callosum. This is a mass of nerve fibres, which passes messages from one side of the brain to the other to co-ordinate our actions and cognitions. We need this, because the two halves work together but have slightly different functions. In general, the left cerebral hemisphere controls the right side of the body while the right cerebral hemisphere controls the left side of the body. There are some other differences between them (although not as many as some people have claimed), which we'll look at in Chapter 2.

The two cerebral hemispheres are physically very similar in structure, so the way that we name the different areas is the same for both sides of the brain. The surface, as we've seen, is covered by deep grooves with rounded areas between them. Each groove is known as a **sulcus** and the mound between the grooves is known as a **gyrus**. (The plurals of these are sulci and gyri, in case you come across these terms.) The deep sulcus, which divides the two halves of the cerebrum, is known as the medial sulcus. While the medial sulcus is the dividing line between the left and right cerebral hemispheres, it doesn't separate them entirely – as we've seen, they are joined by the corpus callosum – but the join is so deeply buried between the two that it cannot be seen from the surface.

On each hemisphere are two particularly long and deep sulci. These divide each cerebral hemisphere into four 'lobes'. The lateral sulcus, sometimes also called the lateral **fissure**, runs along the side of the brain, and the area of the brain below it is called the temporal lobe. It has many functions, as we'll see throughout this book; and one of its distinctive ones is in processing the sounds that we hear. The other major sulcus is the central sulcus, or central fissure, which runs across the top of the brain. It separates the front part of the brain, known as the frontal lobe, from the area behind it, which is known as the parietal lobe. The frontal lobe is particularly concerned with decision-making, planning and movement, while the parietal lobe integrates various kinds of sensory information. The fourth lobe of the brain, the occipital lobe, is not defined by a sulcus as the other three are, but it is the area at the very back of the brain and it is distinctively concerned with vision.

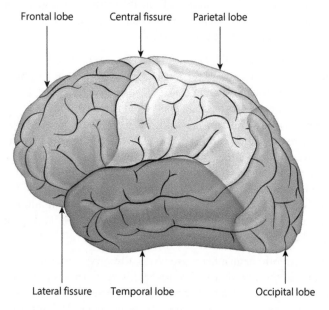

Figure 1.6 Cortical lobes and fissures

Some of the most interesting parts of the cerebrum, though, are tucked right underneath the lobes, where they fold in on themselves. For example, under the cerebrum, but still part of it, is a thin but widespread layer of cells that links together many different areas of the cerebral cortex and connects with many of the structures in the limbic system. This is called the **claustrum**, which some researchers believe is crucial to consciousness and what we experience as our connected awareness. We'll look at this again in Chapter 13. Then there is the **insula**, an area of the cortex folded deep inside the lateral fissure. This, too, is involved in consciousness, and also with social perceptions such as empathy, compassion, self-awareness and emotional experiences. Not only that, but the insula has strong connections with our systems for control of movement and cognition – that is, thinking and memory.

The brain, then, is a complex structure, and exploring how it works, and how its different structures connect with one another, is one of the most exciting aspects of modern science. Since neuroscientists are making new discoveries all the time, we are unable to cover them all here, but I hope this book describes enough of them for you to get an idea of what is where, and how your brain makes you into the person you are.

Focus points

* The brain began as an extension of a simple tube linking a primitive nervous system. It became increasingly complex as animals evolved.
* The parts of the brain closest to the spinal cord deal with essential life-maintaining processes, like breathing, heartbeat, digestion and alertness.
* Larger subcortical structures include the thalamus and the cerebellum. The thalamus co-ordinates sensory information while the cerebellum co-ordinates movement.
* The limbic system is a collection of small structures that includes the amygdala, the hippocampus and the basal ganglia, which are important in emotions, memory and learning.
* The largest part of the human brain is the cerebrum, which is divided into two halves that cover almost all of the rest of the brain, and are wrinkled and folded to increase their surface area.

Next step

In the next chapter, we'll look more closely at how the different parts of the brain work and what brain studies and brain scanning reveal.

2

How does the brain work?

In this chapter you will learn:

- ► *how chemicals and electricity allow brain cells to communicate*
- ► *how we learn and the nature of neural plasticity*
- ► *what brain lateralization is*
- ► *what studies of the brain have shown*
- ► *what brain scanning reveals.*

Brain cells

In Chapter 1 we saw how different parts of the brain do different things. This chapter describes how these different parts send messages to one another, combining their actions to produce the living, breathing human being that is ourselves. Through a combination of chemicals and electricity, the various parts of the brain communicate with each other and with the rest of the body. It might be a good idea, though, to start by looking at the cells that make up the brain.

The surface of the brain is composed of grey matter – sometimes researchers just call it 'the gray' – but underneath it is a mass of white matter. This white matter consists of nerve fibres passing messages around from one area of the brain to another, which is how everything is interconnected. The nerve fibres are white because they are myelinated (see below).

Most of the cells that make up the grey matter are **interneurones,** sometimes called connector neurones. Their main purpose is to make connections between nerve cells, so their structure is relatively simple. An interneurone consists of a cell body with many projections extending outwards to form branches, or dendrites. Sometimes there is a longer projection from the cell body, known as an axon, and the dendrites are found at the end of that. Each dendrite ends in a small lump, called a synaptic knob, which makes a connection with another neurone.

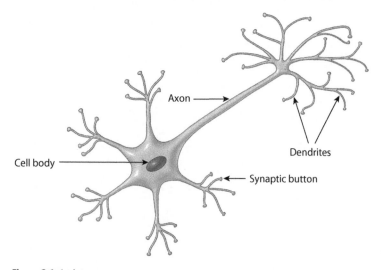

Figure 2.1 An interneurone

There are other types of neurone in the nervous system. Sensory neurones pick up signals at the sense receptors – the eyes, the skin, the nose, the ears and so on – and pass them on to the brain. These neurones have specialized receptor areas at one end, which pass the signal along to the cell body. From there, the message travels along an axon to the relevant parts of the brain. Other neurones called motor neurones take messages from the brain down to the muscles, allowing us to move. Figure 2.1 shows the general structure of an interneurone, but all neurones end in dendrites, with synaptic knobs on the end.

Besides neurones, there are a number of other cells in the brain, which go by the general name of **glial cells**. Their main function is to support the neurones by holding them in place and providing them with oxygen and nutrients. They also remove dead nerve tissue and toxic substances, they help to insulate the neurones from one another, and they can be important in stimulating cell growth.

So the brain is a closely packed, dense mass of cells, but, as we saw in Chapter 1, it still has recognizable structures. It also has some quite large spaces – fluid-filled areas known as **ventricles**. These are inside the brain but linked to the spinal canal: a relic of the early neural tube. The ventricles are filled with cerebrospinal fluid, which is a clear liquid that provides nutrients and immune support to the brain structures and clears away waste products. The ventricles can also act as a shock absorber, like an airbag, protecting some of the most vital brain structures against impacts.

Chemicals and electricity

Brain cells work, essentially, by using chemicals to generate electricity. Like all living cells, they have a slightly different electrical field than their surroundings. Inside the cell there's a slightly higher concentration of potassium ions, which have a negative electrical charge. Usually, the membrane surrounding the cells stops other chemical ions from passing through. But if it is stimulated in the right way, it changes its structure, letting sodium ions through. These have a positive electrical charge, and the exchange of positive and negative ions generates a sudden burst of electricity in the cell.

We call this sudden burst of electricity an **electrical impulse**. Electrical impulses travel around the brain by moving along the extended 'arms' (axons) of the neurones. Sometimes they travel relatively slowly as each impulse changes the next part of the cell membrane, depolarizing it so that sodium ions can get in to renew the electrical impulse. But that's a relatively slow and gradual way of getting the message along. Neurones that need to send messages quickly have a different structure: their axons are covered by white matter, which helps the message to move more quickly.

The white matter in these cells is a fatty coating known as a **myelin sheath**. It's made by special cells called **Schwann cells** wrapping themselves around the axon, with very small gaps where the cell membrane is exposed to its surroundings. Each Schwann cell insulates the axon, preventing the exchange of positive and negative ions. This means that the electrical impulse can only be renewed at the gaps between the Schwann cells, and the impulse has to travel along the axon in large jumps. That's much faster, and that's what the white matter of the brain is all about. It consists of billions of myelinated nerve fibres, buzzing with electrical messages flying from one part of the brain to another.

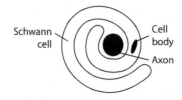

Figure 2.2 A Schwann cell

That's how electrical messages pass along the neurones. But how do they pass from one neurone to another? This takes us back to the synaptic knob mentioned earlier. The connection point between two neurones is known as the **synapse** – a gap between two neurones. Each synaptic knob faces a receptor site across the gap, at the next neurone. Synaptic knobs contain small 'pockets', or vesicles, which are filled with a special chemical called a **neurotransmitter**. When the electrical impulse reaches the synaptic knob, the vesicles open and spill their neurotransmitter into the gap. It is then picked up at the receptor site on the dendrite of the next neurone, changing the electrical polarity of the neural membrane. Any one neurone will always contain the same neurotransmitter in its vesicles, but there are many different transmitter chemicals used in the nervous system. We'll look at the action of some of these in Chapter 13, when we explore drugs and consciousness.

The influence of the chemicals from one single synapse wouldn't be enough to make another neurone react. But if enough synapses are stimulated, the cell membrane of the next neurone changes. Some synapses make the next cell more likely to fire, and these are known as excitatory synapses, because they excite and stimulate the neurone. Other synapses make the neurone that is receiving the message less likely to fire, and they are known as inhibitory synapses because they inhibit firing. The combination of excitatory and inhibitory synapses creates pathways through the brain, directing impulses towards some brain areas and away from others.

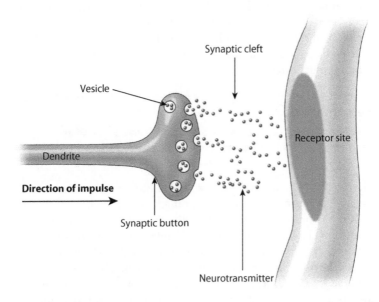

Figure 2.3 The synapse

Learning and neural plasticity

Throughout the cerebral cortex – indeed, throughout the brain – there are **neural pathways** channelling electrical impulses from one area to another. These pathways are partly inherited but they are also strongly shaped by our experiences. When a baby is born, its brain has far more connections between neurones than at any other time; just about every possible synapse is 'live'. But over the first three years of life, these connections are pruned down: connections that are used become stronger, while unused connections die away. That still leaves us with billions of synaptic connections, of course; it's only a gradual pruning, not a severe one.

Why does this happen? It's all to do with how adaptable we are as a species. Human beings – or at least human societies – are found in all sorts of different environments, from the frozen Arctic to baking deserts, to jungles, and even in mid-continental environments where temperatures can vary from +40 °C to –40 °C during the course of a year. We can survive in all these environments, even though the survival skills we need for them are very different. But we don't survive by adapting physically; we do it by learning. From the moment it is born – or from even before birth – a human infant is learning about its world. After birth, its main orientation is to learn from the other people around it, and that learning can take many forms. By the age of three, a human child has pretty well learned to cope with its physical world, at least. Coping with the social world is more complicated!

Remember this

There's more about how we learn and adapt our behaviour to different environments in my Teach Yourself book *Understand Psychology*.

Humans have an incredible capacity for learning, and that's what the human brain is all about. Learning happens as the nerve cells in the brain strengthen particular synapses, weaken others, and channel nerve impulses through unfamiliar pathways to produce a given effect. Our brain cells respond to the demands of new learning in two ways. One way is by building up the synapse – or, at least, growing the capacity of the synaptic knob and its associated receptor site – so that more neurotransmitter is released and picked up, making a stronger message. The other way is through myelination. A neurone that fires only occasionally doesn't generally develop a myelin sheath, but Schwann cells are attracted to active neurones. So if a group of cells continue to be stimulated, as they will when we are learning something new, Schwann cells begin to wrap themselves round their axons. As we've seen, myelination helps the message to pass along the neurone more quickly, so we find it easier to do the thing we are learning.

A lot of our learning happens when we are infants, but we continue to learn throughout our lives. This means that we are constantly placing demands on our brain cells, encouraging them to form new connections. For example, we are not born with the ability to read; neither did we evolve that capacity. In most human cultures, reading, if it existed at all, was a limited skill, restricted to only a few individuals. Yet we are all able to learn how to do it. It takes sustained effort and a lot of experience with the written word, yet most people can read fluently by the time they are ten or twelve years old. Some people may take longer, depending on their experiences and their motivation to put in the effort required.

The cerebral cortex consists of many groups of neurones, known as **nuclei**. As we learn to read, certain nuclei in the brain are stimulated and begin to develop new connections. It may be that some of these nuclei evolved originally to be sensitive to natural signs or symbols – to warn us that red might be dangerous, for example, since it's the colour of blood. But however they evolved, given the right kind of experience these nuclei will adapt to deal with other kinds of symbols as well. If that experience is repeated exposure to the written word – and, most importantly, to its meaning – the nuclei gradually adapt to allow us to read fluently. The more experience we have of reading, the more that group of neurones develops, until, ultimately, we end up with an area of the brain specifically adapted for decoding written words.

The same thing is happening all over the brain, as we learn to deal with other demands made by our environments and experiences. It's not random: certain areas develop in certain ways because those nuclei are predisposed to deal with that kind of information. Much of that has been shaped by our evolutionary history, so we find similar areas in the brains of other mammals. But how far we take it and how sophisticated our skills become have everything to do with our human ability to learn. Our evolution has demanded that we learn new skills to cope with ever-changing environments, so we have evolved the brain structures that allow us to do this.

The ability of brain cells to adapt is known as **neural plasticity**, and it continues throughout our lives. It used to be thought that the brain would adapt only up to the age of puberty and that after that the function of our brain cells was largely fixed. Now we know this isn't the case. Yes, children do recover from brain damage more easily than adults, sometimes even by regrowing areas of the brain, but adults are also able to recover from many types of damage to the brain, by rechannelling neural impulses so that they form new pathways. We know, too, that neurones can continue to grow and develop throughout our lives, as long as they receive the cognitive or physical demands needed to stimulate that growth.

Case study: Growing a brain

The process of neural growth is graphically illustrated in the story of Noah Wall, 'the boy who grew a brain'. Noah was born with spina bifida and hydrocephaly (water on the brain), which left very little space for brain tissue. At birth, he had almost no cerebrum: the space in his skull was taken up by cerebrospinal fluid under considerable pressure. Most infants born with this condition do not survive, but Noah was lucky in having very loving and determined parents, who gave him intensive stimulation and activity throughout his waking hours. Surgical interventions released the pressure from the fluid, and despite starting off at birth with very little brain, Noah's brain tissue responded to the demands of his enriched environment, and his cerebral hemispheres grew and developed. By the age of five they were close to normal size, and he was, to all intents and purposes, a normal little boy.

The human brain retains its ability to adapt to new physical circumstances, too. A study of astronauts showed how exposure to micro-gravity can produce changes in the brain. Koppelmans et al. (2016) used MRI scans to compare the brains of shuttle crews and International Space Station astronauts before and after their time in space. They found that the astronauts' brains had developed more grey matter around the areas particularly concerned with lower limb movement. The longer they had spent in space, the more this example of neural plasticity was evident. The lower limbs are particularly important on Earth, in gravity-bound movement, but are much less important in the micro-gravity of space, so the researchers suggested that the neural changes came from the brain trying to adjust to this difference. A similar, though not identical, result came from comparisons with people experiencing prolonged bed rest.

Other studies have shown how the brain is able to reorganize itself after the damage produced by a stroke – the sudden interruption of the blood supply to an area of the brain. The lack of oxygen causes crucial nerve cells to die, and normal functions such as movement or speech to become impaired. We know from clinical experience that people are able to recover from this, often getting almost full functioning back – but they do have to put a lot of effort in if they are going to manage it. The nerve cells in the brain respond to those efforts by reorganizing themselves, bypassing damaged areas and developing new pathways to achieve the action or ability being demanded of the body.

Even people who have lost whole areas of the brain can sometimes regain functions. In Chapter 10 we'll be looking at the language areas of the brain, which are mainly – though not always – found in the left hemisphere. Damage to these parts of the left hemisphere can seriously interfere with a person's ability to use language – to speak or to form words, or even to understand them. But in 1980 Gooch reported on the results of a dramatic operation. A few clinical patients had such severe damage to the left hemisphere that surgeons decided to remove that half of the brain altogether. Before the operation, they had been entirely unable to use language, but when the damaged hemisphere was completely removed, they began to recover: to be able to speak, understand, and even to remember the words of old songs. The right side of their brains had taken over the language functions that had previously been located on the left. This level of plasticity was entirely unsuspected, and it showed how making simplistic models of how the brain works is generally a mistake. It's always more complicated than we first think!

Brain lateralization

Gooch's report challenged the idea that language is located only in the left hemisphere, and showed how adaptable our brain hemispheres can be. As a general rule, though, there is a certain amount of lateralization in the brain: one side of the brain does one thing while the other side does the other. For example, the left side of the brain controls the right side of the body and the right side of the brain controls the left side of the body. Instructions from the right side of the brain to move your hand, for example, will result in your left hand moving, and vice versa.

The exceptions to this are the senses located in the head: your eyes and ears both have crossover points, so that both sides of the brain receive information from each eye or ear. The ears need this because assessing the differences in the sound coming to each ear is an important cue that tells us where sounds are coming from. In the eyes, the crossover means that information coming to the left side of your eye – that is, from things to the right of your vision – go to the left side of the brain, while information coming into the right side of your eye goes to the right side of the brain. So each eye can see the whole visual field, but the brain can also compare the two images.

Researchers have been able to use this to explore how different sides of the brain work. Using a screen to hide one part of the visual field, they have shown how the left side of the brain is able to understand written instructions, while the right side doesn't usually read but can understand other types of meaning. There's a bit of crossover too: for example, if you showed the right side of the brain the word 'key', the person might not be able to say what the writing said, but they might be able to pick out a key from a tray of objects. So there would be some grasp of meaning on that side of the brain, but not confident reading.

Most people have a preferred hand, and sportspeople know that we also often have a preferred foot, although the two are not necessarily the same. Most people are right-handed, but many people are left-handed. Right-handed people usually have a clearly dominant left hemisphere, but left-handers show more balance between the hemispheres, with a lot of activity on the left side as well as the right. They also show more variability in which hemisphere their main language use is located – right, left or, as some people show, equal activity between the two. Right-handers are much more likely to be left-hemisphere dominant for language. But nobody has ever shown a link between hemisphere dominance and cognitive ability: whichever side of the brain you prefer to use, your abilities and skills have equal potential.

Key idea

Are you right-handed, left-handed, or ambidextrous? Most people see themselves as right-handed, because they write with their right hands. But just about a quarter of those people are actually mixed-handed: they may use their right hands for complex tasks like writing but use either hand for other ordinary functions. It all depends on how we define handedness. In the same way, estimates of how many people in the population are left-handed vary from 4 per cent to 30 per cent, depending on how strictly left-handedness is being defined.

There are a number of common myths about the right and left hemispheres of the brain. You may hear, for example, that the right brain is more creative while the left brain is more analytical, or that the right brain is mystical while the left brain is materialistic. These are largely nonsense, being exaggerations from some much more precise scientific observations. Those scientific observations include the way that spatial problems like those involving diagrams tend to be processed more in the right cerebral hemisphere, while arithmetic problems involving sums or counting are generally processed more by the left side of the brain.

This observation, which in any case is only a trend and doesn't apply to everyone, led to a common myth. The myth went something like this: spatial awareness = drawing = art = artistic personality = creativity; while arithmetic = numbers = counting = practical personality = materialism. But there's no evidence for those conclusions; in fact, we know that creativity, drawing as it does on skills, memories, abilities and imagination, involves both hemispheres of the brain. Similarly, mathematicians use the right hemisphere as well as the left when processing mathematical problems. While there is some lateralization of our brain functions, it's important to remember that the two halves of the brain generally complement one another. They are not opposites: they work together to give us our experiences. One side of the brain may analyse words for meaning, while the other analyses the spoken nuances of tone of voice and timings – and together they allow us to make sense of what people are saying to us.

There have been some studies of people who have had the band of fibres known as the corpus callosum cut through, in an attempt to control severe epilepsy, which starts in one hemisphere but then spreads across the brain. When these people were tested, researchers found that the two halves of the brain could each operate independently, and that each had some of the abilities normally reserved for the other. Language functions were on the left, for example, but the right hemisphere also showed the ability to read simple words. Similarly, the right hemisphere was better at processing pictures, but the left also showed some of that ability. But what was particularly interesting were the accounts given by some of the people concerned. One woman, for example, described how her left hand might pick out a dress from her wardrobe even though she was not aware of making that choice, and had consciously been thinking about a quite different outfit.

Studying the brain

The split-brain studies took place in the 1960s, and they are a good illustration of how limited brain research was at that time. Researchers generally relied on surgery, animal studies or studying people's brains after they were dead. Since the brain is encased in a hard box (the skull), we can't see inside it while it is still active, and even if we could there would be little to see: brain cells work by electricity, which can only be detected using specialized equipment. Scientists who wanted to find out how the brain worked had either to take measurements from the outside or to look at the brains of people who had suffered from injuries to particular parts of the brain.

Curiosity about the brain goes back a long way, and the inability to do effective research didn't stop people from developing theories. One common belief in the eighteenth and nineteenth centuries was that, if a mental 'faculty' was well developed, it would cause that part of the brain to grow. That brain growth would push against the skull, resulting in bumps across the head which could be detected from outside. This was known as **phrenology**, and it was a popular theory for many decades. Although it became a very well-developed 'science', there is no real evidence for phrenology and its practice gradually fell into disuse.

MEASURING ELECTRICAL ACTIVITY

There are other ways of taking measurements from the outside, though, and an early method emerged as scientists realized that nerve cells worked through electricity. Electro-encephalographs, or **EEGs**, are readings of the overall electrical activity of the brain. They are measured by taking readings of the electrical field at various points across the scalp, and they can tell us quite a lot. Chapter 13 describes how psychologists can identify different levels of sleep using EEGs; and how general patterns of activity in the brain can tell us about conscious mental states.

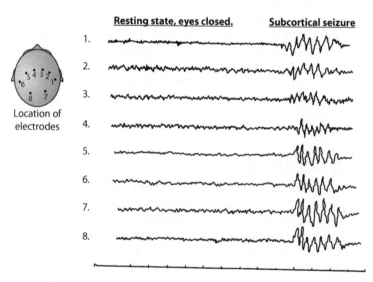

Figure 2.4 An EEG chart

The early EEGs were also able to show what was happening when someone experienced an epileptic fit – something that historically had been regarded as quite mysterious. EEGs showed, for example, that epileptic seizures tend to begin in the temporal lobe on the left side of the brain and then spread outwards, and this was the reason for the split-brain operations mentioned earlier. The idea was that cutting the corpus callosum would restrict the electrical activity to one side of the brain, leaving the other half operating normally. This procedure was used only for very severe seizures, though. EEGs also showed that there are many different degrees of epilepsy, some of which are barely visible to an observer but nonetheless affect the person experiencing it.

As electrical monitoring became more efficient, new techniques developed. One of these is the **evoked potential response,** which is a measure of how an area of the brain responds to electrical stimulation. This has helped neuroscientists to identify some of the major pathways and connections in the brain. Other techniques included the use of **microelectrodes,** which are so small that they are able to direct or stimulate a single neurone. The key studies that showed us how the visual cortex decodes shapes and images used this approach. Eventually, the findings from years of painstaking microscopic research resulted in the Nobel Prize for the two main researchers concerned: Hubel and Wiesel. We will learn more about their findings in Chapter 3.

Hubel and Wiesel's research began in the 1960s, and that decade also saw another major advance in understanding how the brain works, which was the identification of specific neurotransmitters. As we saw earlier, the electrical activity of the brain is generated by the action of chemicals, which pass messages from one nerve cell, or neurone, to the next. Discovering what some of these chemicals were meant that scientists were able to trace neurochemical pathways in the brain. We will see in other chapters how much this has helped to clarify our understanding of how the brain works.

Many of these detailed studies involved animal research. Studies of human beings tended to be limited to the external observations discussed earlier, or to clinical studies of people who had experienced injuries to the brain. Clinicians looked at where these injuries were, and attempted to correlate this with the resulting psychological deficits or changes in personality. Sometimes the results were fairly clear: as we shall see in Chapter 10, Broca and Wernicke were able to identify key language areas back in the nineteenth century, by studying people who had specific language deficits and correlating their symptoms with post-mortem studies of their brains which showed damage to particular areas.

When brain studies concerned more subtle changes to areas such as personality, things were rather different. This is mainly because all comparisons had to be made retrospectively – that is, by comparing how someone was after their injury with how they thought they had been before. The problem with this is that we all have many different states of mind, and it is easy to attribute a mental or personality characteristic to an accident, when really it was there all the time but unnoticed. For example, old people often remark on how their memory is failing, but comparisons with young people show that they experience just as many memory lapses as older people, if not more. The difference is that younger people don't especially notice them, while older people notice and worry about every instance of forgetting, because they attribute it to ageing. Really, though, it's been like that all their lives.

The same kind of thing can happen after an accident causing brain injury. We may become newly aware of things about ourselves that were actually the same before, but which we hadn't noticed. So we think they are new and attribute them to the brain injury. This doesn't mean that injury to the brain has no effects, but it is very difficult to tease out exactly what those effects are because we don't document or record every aspect of our normal experience.

BRAIN SCANNING

The real breakthrough in brain research came in the 1980s with the development of scanning. Brain scans made it possible to study the active brain for the first time. Instead of having to rely on injuries or animal studies, we can see the brain actually working in normal, healthy people, and this has really opened up our understanding of what is going on. **Neuroimaging** gives us a picture of the brain, showing which parts are active at any given time and which parts respond to different types of stimulus.

There are several types of brain scan. One group of scans uses the interaction between electricity and magnetism to show how the brain works, and the most useful of these are MRI scans. MRI stands for **magnetic resonance imaging,** and this method uses the way that water molecules in brain cells have small magnetic fields, which are slightly different if the cell is active than when it is quiet. The MRI scanner produces a succession of electromagnetic waves, a bit like radio waves, and active brain cells respond to this. The scanner detects and records all these responses, building up an image of the electrical activity happening in the brain at the time.

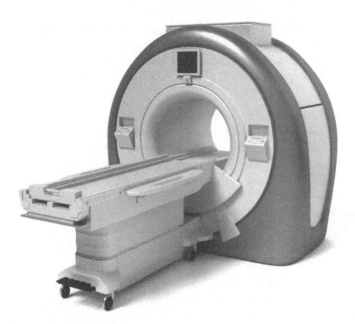

Figure 2.5 An MRI scanner

There are different ways of using MRI scanning. The most common in neuropsychology is **fMRI**, functional MRI, which explores brain activity in relation to specific functions. For example, since a whole MRI brain scan takes only seconds, researchers are able to explore what is happening in the brain while people are actively thinking, such as when they are reading, remembering or solving a puzzle. A succession of scans builds up a picture of changing brain activity during the course of the task.

Another way that MRI scans are used is known as **efMRI**, which stands for 'event-related functional magnetic resonance imaging'. This type of scan compares the patterns of electrical activity produced by two or more different events – for example, looking at the brain activity involved when someone gives a correct answer on a memory test and comparing it with the brain activity shown when they give an incorrect one.

Other types of brain scan include **PET scans**, which trace the distribution of a small amount of radioactive chemical that has been put into the blood supply and taken up by the brain. Active brain cells use more blood than passive ones, as the neurones replenish their nutrients after firing, so the amount of blood used shows which brain areas are more active. In a classic study, for example, Tulving (1989) used radioactive isotopes of gold to trace memory functioning while people recalled episodes of their holidays. The gold isotopes did not stay in the system for long, but were able to indicate which areas of the brain were active at any given time. Medical PET scans use more prosaic substances, but the principle is still the same.

CAT scans – short for computed axial tomography – scan the brain by taking a series of X-ray or ultrasound images in slices, and combining those slices to form a 3D image. They compare different levels of density in the brain. Grey matter, for example, is less dense than white matter, so it looks different in a CAT scan, and so do tumours and blood clots. The image is static, but it does allow a researcher to identify abnormal structures or growths, and comparing CAT scans over time can highlight large-scale developments, like the recovery from damage caused by a stroke or other injury.

In some ways, EEGs can be seen as the first type of brain scan, showing as they do the overall levels of brain activity in different parts of the brain. They are taken by attaching electrodes across the head; nowadays researchers tend to use a net, rather than sticking the electrodes directly to the scalp as they used to do. The electrodes are sensitive to the electrical emissions coming from the brain. Advances in technology mean that EEGs have become much more sensitive than they used to be: one classic description of the old-style

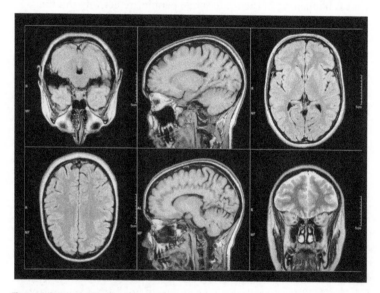

Figure 2.6 A typical brain scan

EEG as a method of research was that it was like standing outside a factory and trying to work out what was going on inside from the noises coming through the window! But even so, they did manage to identify general patterns of activity, such as the alpha, beta and delta waves associated with different mental states – alpha with relaxation, beta with alertness and wide-awake states, and delta with deep sleep. We'll be looking at this more closely in Chapter 13.

There are also a number of variations of this type of approach, such as **MEG scans** (magneto-encephalography), which use SQUID (superconducting quantum interference devices) to detect changes in the magnetic activity of the brain. These devices are extremely sensitive and can detect minute adjustments to the magnetic field around the brain, so they can home in on particular areas even from outside the scalp. ERP (event-related potential) measures are able to identify changes in the electrical activity of a region of the brain in response to a stimulus or cognitive event and, as we saw earlier, an early variation in EEG research was the use of evoked potentials, where a stimulus, like a sound, would be given and the brain response measured.

Transcranial magnetic stimulation (TMS) is a particularly interesting way of studying the brain. It involves administering a short burst of magnetic stimulation. This disrupts the brain's activity and interferes with any processing it is doing at the time, but has no lasting effects. TMS is relatively easy to administer because it doesn't involve the whole scalp. It is applied just to the outside of the head in a particular area. Applying TMS to the side of the head near the junction of the parietal and temporal lobes, for instance, can seriously (but temporarily) interrupt someone's language abilities, having a dramatic effect on the performance of a task like reciting a poem or reading aloud. Transcranial direct current stimulation (**tDCS**) is a very similar process: an electrical coil is held directly over the scalp, producing a 'virtual lesion' interfering with brain functioning. It can work in two ways: cathodal tDCS decreases the level of brain activity, interfering with performance, while anodal tDCS increases the level of activity, enhancing performance of particular tasks.

The brain has other surprises, too. Brain scans have shown us how common **mirror neurones** are in the brain. The first mirror neurones to be discovered were in our movement systems, and they showed that our brain activity doesn't just reflect our own actions. We show similar patterns of activity in the same parts of the brain when we are watching other people do things, as we show when we are doing them ourselves. It happens only if we are paying attention, of course, but if we were watching, say, a tightrope walker and imagining what it would be like, some of our own brain cells involved in balance and walking activity would also become active.

Since that first discovery, researchers have found mirror systems in many parts of the brain, and particularly in those areas involved with social interaction – speech and conversation, social memory and so on. When we are with or watching others, our brain activity is structured to empathize with them to some extent. We are even more strongly social than we thought! As you read through this book, you'll come across mirror neurones quite often.

Focus points

* The brain is made up of nerve cells called neurones, which make connections with one another and carry messages around the brain and to the rest of the body.
* The brain's messages take the form of electrical impulses, which jump from one neurone to another using special chemicals called neurotransmitters.
* Neural connections develop as we learn new things, and the brain is able to adapt to damage even in adulthood. This is known as neural plasticity.
* The left side of the brain controls the right side of the body, and vice versa. Some other functions are also lateralized on one side of the brain or another, but not as many as popular myths imply.
* Early brain research had to rely on damage, animal studies or EEGs, but brain scanning allows us to study the brain while it is working. There are many different types of brain scan, but PET, CAT and MRI scans are the most commonly used.

Next step

In these first two chapters we have seen the general picture of what the brain is like and roughly how it works. In the next chapter we'll begin our specific exploration of how different parts of the brain achieve our experience of what it is to be a human being, starting with how we see the world.

3

How do you know what you're looking at?

In this chapter you will learn:

▶ *how the visual system works*
▶ *about an aspect of vision known as blindsight*
▶ *how we see and identify objects*
▶ *how we detect movement*
▶ *how we see people.*

Our vision is amazing: we can see things at a distance or close up, in colour or monochrome, moving or still, in near-darkness or in bright light. While some animals have more acute vision than we do, or can detect the further ranges of the electromagnetic spectrum, our visual system gives us a wealth of information and is ideal for an adaptable animal like ourselves. This is why vision is the most important of all of our senses – so much so that we have developed elaborate ways of helping people with restricted sight or blindness to cope with their disability. We don't bother at all about people with restricted olfaction, who can't smell clearly, which tells us just how important we think vision is.

How does vision work? It's all based on light, of course, and it's all about the information that light carries to our eyes. The brain, as we've seen, works on electricity, so we have developed elaborate structures to convert light information to electricity. It begins with our eyes, which are organized so as to collect light and project it to the retina – a layer of cells at the back of the eyeball. These cells contain chemicals that react to light by generating a tiny electrical impulse. The impulses then pass from one nerve cell to another, and most of them eventually arrive at the very back of the brain, in the part that we call the visual cortex. This is a large area of the brain and it's the source of our conscious visual experience. But a lot of sorting out of the information happens on the way.

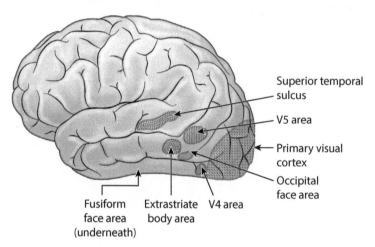

Superior temporal sulcus

V5 area

Primary visual cortex

Occipital face area

Fusiform face area (underneath)

Extrastriate body area

V4 area

Figure 3.1 Visual areas of the brain

Vision and blindsight

Visual information can take several different routes between the eyes and the brain, and many of them are quite ancient in evolutionary terms. As we evolved more complex systems, we also kept the earlier ones, and since human beings are quite highly evolved, we have quite a few different ways in which light information can influence the brain. This can sometimes produce interesting effects.

Have you ever been busy doing something and then suddenly become alert? Something happened, but you only work out later what it was that attracted your attention. That's because of one of these ancient pathways. We saw in Chapter 1 how the two superior colliculi of the midbrain have a direct connection with our alarm and alertness systems. They receive information from our sense receptors, so if something sudden happens, we react immediately. A sudden change in light or sound produces an automatic reaction without any thought being involved. The thinking comes later.

If you've ever stayed awake all night, you'll probably remember how much more awake you felt as dawn happened and you saw daylight. We are more alert during the daytime than we are during the hours of darkness, and this is because there is a direct neural connection between the retina and the hypothalamus. It provides information about day and night, and that helps the hypothalamus to regulate our biological rhythms. Artificial light can interfere with this, of course, but we still have the basic biological rhythms, and they respond strongly to light even in people who are not consciously able to see.

Have you noticed how moving things catch your eye? If you're looking at a scene and you see something moving, your eye is immediately drawn to it. This is another useful survival mechanism, helping us to notice potential predators or other people. Our visual system is immediately attracted to movement because another pathway bypasses the usual channels to the main visual cortex, and goes directly from the thalamus to the V5 area of the visual cortex. This is the part of the visual area concerned with visual motion, and the direct pathway means that we can detect movement around us without being fully aware of what we are seeing.

Altogether, researchers have identified about ten different ways that information can travel from the retina to different parts of the brain. These discoveries have allowed us to explain one of the more puzzling aspects of human vision, known as **blindsight**. Back in 1972 Weiskrantz studied patients who were apparently quite blind, but who nonetheless would react to visual stimuli. They might, for example, be able to point to a moving object or even duck if something were coming towards them, although they had no knowledge of what they had seen. They felt as though they were just guessing, even when they responded accurately in laboratory tests. These people – and many others who have been studied since – had damage to their visual cortex, and so they were unable to process visual information consciously. They were blind, but the other, more ancient aspects of their visual system were still functioning.

There are other odd forms of blindness, too. Some people develop specific problems with their vision, usually as a result of an infection in the brain. Perhaps the most common of these is categorical blindness, in which the person becomes completely unable to identify things in a particular category. Often, this relates to animals. They can recognize anything around them except a dog or cat, say, or any other kind of animal. When they see those they are completely baffled, unable to work out what they are. Categorical blindness can work in other ways, too: some people are fine with living things and anything from the natural world, but they cannot recognize artificial, constructed objects like tools or telephones. And some people can recognize animals and objects, but are completely unable to fathom what food is when they are looking at it. They can eat the food but they just can't recognize it when they see it.

Categorical blindness comes from damage further along the visual system, in the areas where the brain makes sense of the images it is receiving. But the types of categories that are disturbed are also linked closely with our evolutionary history. In the early years of our evolution, distinguishing between animals and objects would have been fundamental to survival, and so would being able to identify food. The fact that these categories can become specifically disturbed is because they are so important that they have become 'hard-wired' into our brains: we are far more ready to make those classifications (animal, food, object) than others that are more modern (building, transport, sign).

How we see

What we generally think of as seeing, though, which is the ordinary seeing that we are aware of, uses the main route for visual information, which has been well documented over the years. It begins with light-detecting cells, called **photoreceptors**, in the retina. These are of two kinds: the extremely sensitive rod cells, which detect brightness, and cone cells, which detect colour and only really work in fairly bright conditions. The role of both types of cell is **transduction**: changing light information into the electrical impulses that the brain understands. They manage it because the light bleaches special chemicals in the cell, which changes its electrical potential.

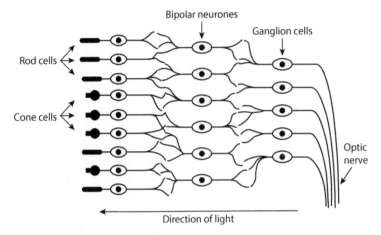

Figure 3.2 The structure of the retina

Once the information has been converted into electrical impulses, it is passed to a second layer of cells in the retina, which are the bipolar neurones. These do the first bit of processing: they respond either to light areas on dark backgrounds or to dark areas on light backgrounds. This primitive type of processing allows us to detect simple features in the environment: a pool or the sea, for example, generally reflects more light than the earth surrounding it. Since the brightest area is usually the sky, at least in the daytime, it's helpful to distinguish bright areas on the ground (or towards the bottom of the **visual field**, if we want to get technical about it).

For many animals, being sensitive to movement is a matter of life and death, and processing movement is basic in the visual system. The third layer of the retina contains the ganglion cells, which detect movement by responding to changes and differences in the visual field. Each ganglion cell has its own receptive field, centred on one bit of the retina and extending outwards from it. Some of the cells react when light falls on the centre but not on the surrounding area, while others work in the opposite way, reacting when light falls on the surrounding area but not on the centre. They also react to any change from that arrangement, which makes them particularly sensitive to movement.

We are much less able to see things that aren't moving. A cat or dog may fail to spot something that is not in motion, and may need to use other senses to detect it. We ourselves are only able to perceive things that are still because our eyes make continual jerks and tremors, known as saccades. This means that the neurones in our eyes are constantly adjusting to slightly different input as if things were moving, even though they are really staying still.

Ganglion cells have long axons, which bunch together at a particular point and form the optic nerve. There's a blind spot where the optic nerve leaves the retina, but you wouldn't know it because the brain fills in the missing information for you. The optic nerve takes the information to the lateral geniculate nuclei of the thalamus, for more processing. On the way, it passes through a crossover point known as the **optic chiasma**. Information from the right-hand side of the retina in each eye goes to the right side of the brain; and information from the left side of each retina goes to the left side of the brain. So both eyes pass messages to both parts of the brain, but information from the left side of the retina goes to the left side of the brain, while the right side of our brain receives information about the left visual field – in other words, information that has been received on the right-hand side of the retina. If that confuses you, take a look at Figure 3.3.

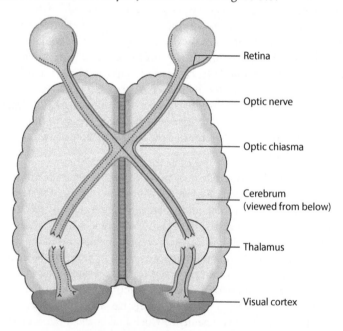

Figure 3.3 Information pathways from the eyes to the brain

Our visual information is sorted again when it reaches the thalamus. The thalamus has six layers. The top four layers respond particularly to detail and colour, while the bottom two layers co-ordinate information about movement. They react to movement and change across large areas of the visual field. So the thalamus brings together and organizes different types of visual information, before passing it on to the visual cortex for conscious processing.

Even before we achieve conscious sight, then, the information from our eyes has been sorted out quite a lot – and it has all had evolutionary value. Distinguishing between light and dark areas, movement and change helps any animal to avoid or react to obstacles and its environment, and to detect potential food items or predators approaching. We've also seen how some of this information goes directly to the more primitive parts of the brain, stimulating immediate reactions.

But that's all very basic. More sophisticated visual processing happens at the back of the cerebrum, in the area known as the visual cortex. The main area is known as the primary visual cortex, or V1 for short, and its role is making sense of all the visual input it is receiving. This area of the brain was first clearly identified as a result of the shell damage experienced by men during the First World War. Researchers found that damage to this part of the cortex reliably produced some kind of blindness, and the more damage there was, the more severe the blindness was. Depending on how much of their cortex had been destroyed, some ex-soldiers were blind only in certain parts of their vision while others were totally blind.

In another early study, the surgeon Wilder Penfield stimulated this part of the brain in patients who were undergoing open brain surgery. People are usually conscious during this type of surgery because the brain doesn't have pain receptors, so Penfield was able to ask them what they experienced. They described a range of visual experiences, such as balloons floating in the sky or country scenes. He showed that the visual cortex wasn't just a mass but that different parts of it did different things; but it took a lot more research to find out exactly how it worked, and that research is still continuing.

The visual cortex also makes connections with other areas, and it seems to use two main pathways for this. The first, known as the **ventral visual stream**, is mainly concerned with identifying objects and things regardless of where they are, so it's often called the 'what' stream. It runs from the visual cortex to the temporal lobes of the cerebrum. The second is the **dorsal visual stream,** which is concerned with locating objects and things regardless of what they actually are, so it's known as the 'where' stream. It runs from the visual cortex to the parietal lobes. Working together, the two visual streams allow us to make sense of our world and act effectively on it. So let's look at how our brain cells work to produce the complex visual world that most of us live in.

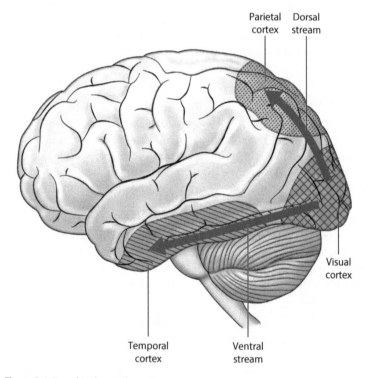

Figure 3.4 Dorsal and ventral visual streams

Seeing things

We've seen how our visual cells react to light and dark, and how this may be a basic survival mechanism. But as human beings our vision is a lot more complex. We see objects, backgrounds, people, colours – and our brain processes all of these in one way or another. How does that happen? One of the most important findings in this area came from work by the Nobel prizewinners Hubel and

Wiesel, who investigated how vision works by making painstaking recordings of the actions of single neurones.

In 1969 they showed how some nerve cells in the primary visual cortex – the V1 area – react to lines that were at particular angles and in only one specific part of the visual field. They called these simple cells. Further investigations showed that these simple cells react to similar inputs from either the left or right eye, and that some of them also respond to particular wavelengths of light – in other words, to specific colours. Effectively, these cells analyse the information arriving at the visual cortex and process its basic features. Then they make connections with what Hubel and Wiesel called complex cells. These combine information from several simple cells, so they will react to a line at a particular angle but in any part of the visual field, or to a line of a particular colour anywhere in the visual field. Complex cells then pass their information to hypercomplex cells, which respond to specific shapes or areas (see Figure 3.5).

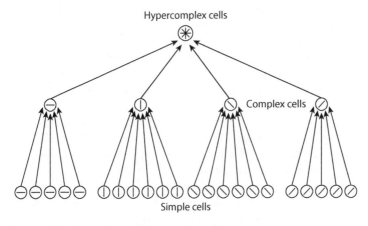

Figure 3.5 Simple, complex and hypercomplex cells

This means that our visual system can distinguish simple shapes and edges, and the boundaries between light and dark areas. According to Marr (1982), this is all we really need in order to perceive the objects around us. If we combine that information from what Marr called the optic array, which is the overall pattern of light reaching the retina, we can identify contours – that is, edges – and regions that are similar. Putting them together will give us the basic structures of a scene, which Marr called the raw primal sketch. Even though it is only vague, it still may give us enough information to detect what an object is, as we can see from the pixelated image in Figure 3.6.

Figure 3.6 The raw primal sketch

But we have more information available to us than just what is in the optic array. Our brains store experiences, and those experiences also help us to make sense of the objects we detect. We know, for example, that objects far away look smaller than if they were closer, and that a near object may cover up part of something that is more distant. We can use the rules of perception identified by the Gestalt psychologists of the first half of the twentieth century, which show how we unconsciously group together bits of visual information in meaningful ways (see my other Teach Yourself book, *Understand Psychology*, if you want to know more about this). As we become familiar with our environments, we learn many 'rules' of perception, and applying these rules, Marr argued, allows us to detect volume, or solidity, in the objects we are looking at. It's not detailed: Marr described it as just a combination of cones and tubes, which he called the 2½D sketch, because it's nearly 3D but not quite. The 'stick-figure image' that results is enough to allow us to identify what type of thing we are looking at.

Some people have speculated that the reason why the artwork of painters like L. S. Lowry strikes such a chord with us is because it taps into our primeval ways of seeing: the stick figures of his paintings are like the 2½D sketch of our primitive decoding of visual information, so they are instantly recognizable. We can easily tell the difference between a cow, a dog or another human, for example (see Figure 3.7), and we can even infer something of the attitudes of the people concerned by the way they are standing.

Figure 3.7 Cow, dog and human stick figures

So we can easily get an idea of what something is, and tell the difference between an animal and a tree, for example. To tell the difference between a dog and a cat, though, might be more difficult. It requires us to draw on our more sophisticated knowledge of the world, using colour, shadows, object contours and, most importantly, our memories, to recognize and identify what we are looking at. But it all starts with the light and dark identified by the bipolar neurones, and developed further by the simple and complex cells in the visual cortex.

JUDGING DISTANCE

Seeing things is all very well but, to survive in the world, we also need to know where those things are, and how close they are to us. The great visual psychologist J. J. Gibson explained how our visual perception is organized in such a way as to help us to move around in the world and to interact effectively with the things we encounter around us. For example, the fact that we have two eyes at the front of our heads means that each of them sees nearly the same thing, but not quite. This lets us compare the image from each eye, and use the difference between them to judge how far away things are. This is one of the reasons why the optic chiasma joins up similar information from the two eyes. When it finally reaches the visual cortex, the information from each eye is organized side by side, in columns, making it easy for the brain to compare the two images.

Why does the brain do this? It's because the slight differences between each eye tell us how far away something is – and that's an important thing to know if, for example, you are jumping from one branch of a tree to another. Arboreal animals – that is, those that live in trees – almost always have frontally mounted eyes, because misjudging distance can be catastrophic. As primates, that's part of our own evolutionary heritage, too. You can identify these differences between the two images quite easily if you hold a pencil out at arm's length. Close one eye and line it up with something in the background. Now close that eye, and open the other. The pencil will now be lined up differently. Do the same thing but holding the pencil closer, and the difference is even larger. That's how your brain can use both eyes to judge distances. It's called **binocular disparity**.

There's more to seeing distance than binocular disparity, though. As we move around in the world, the visual image we receive from our eyes flows and changes along with our movement. That, too, is an important cue to how far away things are, and where we are in relation to them. It's known as the **optical flow,** and we use it more or less unconsciously. Next time you're on a train or a passenger in a car, notice how your surroundings change as you look at them. Things that are far away seem to move with you, in the same direction as you are travelling, while things that are close up go past in the opposite direction. Your visual image flows and changes as you move.

Optical flow works with just one eye, too, and so do other distance cues, like how large something is or how high up it is in our field of vision. Those distance cues are what artists use to paint realistic pictures, and they can produce some interesting visual illusions. (If you're interested, there's more about this in my book *Understand Psychology*.) In real life, though, our brains use movement and optical flow to make sense of what's around us, so illusions are much less common. And people with vision in just one eye can still see distance, even though sometimes it isn't quite as accurate.

SEEING COLOUR
It's also helpful to many animals to be able to perceive colour, which can be another useful aspect of survival. The colour of a fruit, for example, can tell you whether it is ripe and ready to eat, or whether it should be avoided for now. The colours of objects close to us look brighter and more vibrant than the colour of more distant objects, which look greyer or more washed out. Other animals – for example, those that rely more on hunting and less on eating fruit – are less dependent on colour vision. For them, the detection of slight movement is more important. Rod cells are more sensitive to tiny changes, and many animals, like dogs and cats, don't have much colour vision at all.

Humans have a special area on the underside of the cerebrum, just outside the main visual area, which is particularly concerned with seeing colour. It's called the V4 area, and if it is damaged people see the world only in shades of grey. That problem – known as **achromatopsia** – is quite rare, though, because we have a V4 area on each of the cerebral hemispheres, so both of them would have to be damaged. People with damage to just one of these areas report seeing colours as less vibrant, often describing them as washed out or dirty.

What this area does particularly well is **colour constancy,** which is the way we see objects as having the same colours, even under different light conditions. What we see as colour is derived from the wavelengths of the light that our eyes are receiving, but those wavelengths change under different types of light. Something seen under domestic lighting in the evening may be reflecting different wavelengths from the same thing seen in bright sunlight. The V4 area of the brain adjusts to this, so that we see colours as consistent. Colour constancy works so well that we don't notice it at all in everyday life. But the dress example in the following case study shows how powerful it can be when it breaks down.

Case study: The colour of the dress debate

In February 2015 there was a massive, worldwide Internet discussion about the colour of a particular dress. The dress was actually blue and black, but a photograph of it taken under restricted lighting produced an interesting effect. While some people saw it as blue and black, others saw it as white and gold. What was happening was that the people who saw it as blue and black were applying colour constancy, allowing for the difference in illumination, while those who saw it as white and gold were responding directly to the wavelengths of light they were receiving, and interpreting them as if they were seeing them in normal daylight. It was the V4 area of their brains that was doing it, entirely unconsciously. Even when the effect was explained, those people (and I'm one of them) still saw the dress that way, and found it hard to believe that the white and gold colour wasn't true.

SEEING MOVEMENT

Films, TV and video have become fundamental parts of modern life. But they are only possible because of the way that the visual cortex responds to movement. A part of the visual cortex called the V5 area, which is near its outer surface, is the main movement centre of the brain: it co-ordinates our perception, joining up different impressions to create smooth movement. This, of course, is how film, TV and video work. If we are shown a series of lights flashing on and off in sequence, we perceive one single dot moving along a line. Our natural tendency to join up the different images into the perception of continuous movement is the basis of the entire movie industry, and has generated countless billions of revenue throughout the past century. It taps into an earlier survival mechanism, allowing an animal to detect the likely movement of a predator or prey from small glimpses behind bushes or other cover, but it's such a powerful process that we do it entirely without thinking.

You may have seen a demonstration of a set of lights, which seem to be just random blobs while they are still. When they become active, though, it is immediately apparent that they are attached to a person or a group of people. Our brains are particularly sensitive to noticing other people and animals – it's part of our evolutionary heritage – so we are very ready to detect **biological motion**, that is, movement produced by real moving bodies. In a typical study, a person will be fitted with a black suit which has small lights attached to all the joints. If they stand perfectly still against a black background, the lights just appear as random dots. But as soon as they begin walking, running, or moving in other ways, we instantly recognize them as a human being in action.

This happens because the V5 area of the brain has a direct connection with another area of the brain, on the temporal lobes, which is specifically concerned with the movement of bodies and faces. That area is known as the superior temporal sulcus, or STS, and it responds whenever we are seeing bodies in motion. The STS combines both visual and auditory information, and it has connections with our own movement and sensory systems. It also has a mirror neurone system, which helps us to copy or empathize with other people's actions. We'll be coming back to this area in Chapter 7, when we look at movement.

Case study: Damage to the V5 area

Because the brain has two sides, damage to just one side often means that we can continue to function almost normally. But one woman experienced damage to both sides of the brain, in the V5 area. As a result, she was unable to see movement. Instead, what she saw was a series of static images. If she tried to cross a road, one moment she would see a car in the distance, but the next it would be really close; when she tried to pour water into a cup, she didn't see the cup filling. Instead, it would be empty one moment and overflowing the next. Interestingly, though, she was able to detect biological movement from moving sets of lights – at least, to tell that it was a body and not a random group – although she was unable to identify the direction of that movement.

Seeing people

Human beings, as we know, are social through and through, and our brains reflect that sociability. So it shouldn't come as any surprise to find that parts of our brain respond specifically when we see other people. One part becomes active when we are looking at human figures or parts of bodies – either in real life or as line drawings or the kinds of stick figure representations discussed earlier. This is the **extrastriate body area,** or EBA. Its name comes from the fact that it is just outside the main visual cortex, which is sometimes known as the striate cortex because it contains a kind of 'stripe' of darker cells. The extrastriate area seems to be mainly concerned with identifying images of the body, feeding that information to other parts of the brain that are concerned with empathy or emotion.

The EBA isn't concerned with fine detail or with what is happening to the body at the time. It's only interested in representations of the body. Temporary interference with EBA functioning, produced by magnetic stimulation, shows that it becomes active when we are identifying and distinguishing between body parts but not with the actions that those parts may be doing. In one study, people were shown an image of an ordinary hand and an image of a hand pierced by a needle, but the response of the EBA was exactly the same to both images. But it does make distinctions to do with body shape, such as registering whether a body is fat or thin. It has been suggested that some people with anorexia nervosa have damage to this area, which means that they consistently misjudge the size of their own bodies, seeing themselves as much fatter than they really are.

Seeing body parts is one thing, but what happens when we see someone we recognize? There's another part of the brain that is active in seeing bodies, which is closely linked with our social memories and with our memory for people in general. It's a part of the cerebrum that is tucked right underneath the occipital lobe, where it joins the temporal lobe. This area doesn't respond to sketches or stick figures as the EBA does. Instead, it focuses only on whole bodies, and responds differently to people we recognize than to people we don't know. Known as the **fusiform body area,** or FBA, it is located right alongside a similar area, the fusiform face area, which we use for recognizing faces. This is not a coincidence: the two work closely together when we are recognizing people.

SEEING FACES

Being able to identify another individual, and to tell people apart, is essential for any social animal. But identifying faces is a complex task. It involves three elements: firstly, recognizing that a particular pattern of light and shade is actually a face; secondly, identifying that face as belonging to a particular individual; and thirdly, interpreting facial expressions, gaze and other types of facial communication.

The first of these – identifying that something is actually a face – takes place in a part of the visual cortex known as the **occipital face area**. This is just below the EBA and seems to do much the same job, only with faces rather than bodies. The neurones of the occipital face area react when we see faces, or images of faces, but not when we see other objects, or bodies. They seem to focus particularly on the physical aspects of the facial stimulus, and will respond to upside-down faces as well as to faces shown the right way up. It's the first stage in analysing faces, receiving information from the primary visual cortex and passing it on to the other two areas.

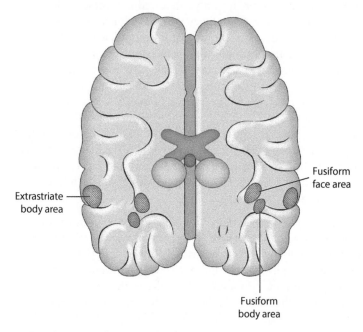

Extrastriate body area

Fusiform face area

Fusiform body area

Figure 3.8 The fusiform body and face areas

The second area, known as the **fusiform face area,** is located underneath the cerebrum, as we've seen, right next to the fusiform body area (see Figure 3.8). The cells in this area respond more to faces than any other type of stimulus, and they are particularly likely to respond to familiar faces. Moreover, they will continue to respond in the same way to the same face, even if that face is shown in different orientations or with different expressions. The fusiform face area has direct connections with the temporal lobe of the cerebrum, which seems to be where the brain stores biographical and personal information, as well as names.

Since there are two fusiform face areas, one on each cerebral hemisphere, it's unusual to find a total inability to recognize any face at all. But some impairment of the ability to recognize faces isn't all that rare. Relatives of people with Alzheimer's are often distressed by the fact that the affected person doesn't recognize them, in some cases as a direct result of nerve cell damage in the fusiform face area. But **prosopagnosia**, as the inability to recognize faces is called, can also be a specific deficit for people who are otherwise quite normal and don't have dementia.

The third area of the brain that is particularly concerned with seeing faces is the STS, or **superior temporal sulcus** – the same area that is involved in perceiving body movement. This area responds to changes in faces: changes of expression, gaze, lip movements and so on. It's important for social and emotional cues, and it receives information from the limbic system and amygdala, among other areas – the areas that deal with emotions. We'll be looking at that more closely in Chapter 8. But these connections mean that this is the area of the brain that links facial expressions with our understanding of other people's emotions.

The STS is also what we use for lip-reading. It brings together information from the visual cortex and the auditory cortex, linking visual input with speech sounds that haven't yet been analysed into words, and it reacts particularly strongly when the auditory input corresponds to the lip movements of the face. That's a powerful aid in conversation and everyday interaction. Training in lip-reading can often help people who are experiencing hearing loss, but we all lip-read to some extent. It's this part of the brain that tells you that something is wrong if you see a film or video where the sound and vision are slightly out of sync. It's less obvious with action sequences, but we really notice when it is faces and spoken words.

We can see the different ways that the fusiform face area and the superior temporal sulcus deal with facial information by looking at the results of a study by Hoffman and Haxby, in 2000. They asked people to make judgements about pictures while using fMRI scans. When people were asked to make judgements about the identity of a particular face, the researchers detected activity in the fusiform face area, but the superior temporal sulcus didn't react. When they were asked to make judgements about eye gaze, though, the FFA didn't particularly respond but the STS was more active.

Our visual system, then, is both complex and social. Some of it is unconscious, some of it is very sophisticated and, as with pretty well all the areas of the brain, researchers are continually finding out more about it. As we look at it, we can see how we have inherited a system that has evolved over time, from the basic decoding of light and dark areas to the detailed recognition of individual people.

Focus points

✻ There are forms of unconscious vision that come from the mechanisms developed as our visual sense evolved. Blindsight is one of these.

✻ In normal, conscious vision, information passes from the eyes to the thalamus, and then to the visual cortex. The optic chiasma allows messages from both eyes to reach the same part of the brain.

✻ Visual information is sorted in the visual cortex to identify basic objects, and combines with experience to tell us about colour and distance.

✻ Being able to detect movement is important to survival, and our brains automatically link separate bits of visual information into continuous movement.

✻ Our social nature is reflected in the way that we have special areas of the brain for seeing people – either bodies or faces.

Next step

In the next chapter, we'll look at how the brain detects and makes sense of the sounds we hear.

4

How do you know what you're hearing?

In this chapter you will learn:

► *the processes involved in hearing*
► *how we hear sounds*
► *how we recognize speech above other sounds*
► *how the brain processes music and rhythm.*

'If a tree falls in the forest and there's nobody there to hear it, does it make a sound?' Philosophers have debated this classic problem through the ages. For a psychologist, though, it's less of a problem. The way we see it is that the falling tree generates vibrations in the air, but it is our brains that detect those vibrations and convert them into what we understand as sounds. Sound is what we hear. If nobody were there to hear it, there would be no sound – except, of course, that other animals would be there and they would be able to hear the sound of the falling tree.

Hearing is our second most important sense. If we can't hear, or if our hearing is impaired in some way, it can affect us a great deal. Hearing loss makes us feel cut off from other people and not fully part of what's going on; and that's why we regard hearing aids, sign language and other forms of support for deaf people as so important. Hearing is a significant part of the way we communicate with other people, and it's also important for keeping track of what is going on around us.

Sounds are the impressions we experience as a result of vibrations carried through the air – or through water if our ears are under water. We humans can recognize a range of sounds, from high-pitched intense sounds, which have high-frequency vibrations, to low, booming sounds with low-frequency vibrations. But, as with light, other animals can detect signals beyond the extreme ends of human hearing. Bats, for example, can produce and detect much higher frequencies than humans can, so a bat in flight might appear to be silent when really it is shrieking out high-frequency calls and listening for their echoes. At the other end of the spectrum, whales produce calls that are so low as to be undetectable to the human ear, but which travel through the water across vast distances, and can be heard by other whales hundreds of miles away. On land, elephants communicate in a similar way, using infrasounds that are too low to be detected by human hearing, but which other elephants can hear over large distances.

Making sense of the changes in air pressure around us isn't just about identifying frequencies, though. We also detect how loud a sound is, and that's conveyed by the amplitude of the pressure waves that reach us. Loud sounds produce waves with high amplitude, while quiet ones have lower amplitude (see Figure 4.1). Sounds are also complex, in that most of the noises we hear consist of several elements, mixing together more than one frequency. 'Pure' sounds, of just one frequency, are very rare, both in nature and in everyday life.

The sound of a flute is probably the closest we come to hearing a pure tone – that is, a sound produced by just one single frequency – but it is the combinations of frequencies which give sounds their identifiable

Low amplitude   High amplitude

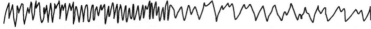

High frequency   Lower frequency

Figure 4.1 Amplitude and frequency

characteristics. We learn these characteristics effortlessly, in some cases almost from birth. We are particularly sensitive to voices, which we can easily distinguish from other sounds. Everyone's voice has a distinctive mixture of frequencies, which means that we can quite easily tell one person's voice from another, and human infants can distinguish the voices of the people looking after them from just a few days old.

The auditory cortex of the brain is also super-sensitive to the timing of sounds – much more so than the visual cortex is to the timing of visual stimuli. We can tell how close together different sounds, or the elements of different sounds, are, even when they are closely timed. This is completely different from our visual system, which, as we've seen, tends to merge together stimuli that arrive rapidly. With sound, the combination of frequency, amplitude and timing of signals means that we're processing quite a lot of information from the vibrations in the air around us.

Case study: James Holman

James Holman was an explorer in the early part of the nineteenth century. He travelled widely across the world, in India, Africa, across Siberia, and through the Australian outback. But Holman never saw any of these places: he had been struck blind by disease as a young adult. Holman trained himself to get around using echolocation – tapping with his cane and listening to the subtle variations in the sound and the echoes that it generated. In this way he could identify objects, gauging their size and whether they were hard or soft, extremely accurately. He wrote several books about his travels, and even argued that his blindness made him a superior traveller because he paid so much more attention to the richness of the information he received from his other senses.

How do we hear?

It starts, of course, with our ears. Have you ever noticed how sounds seem different when we hear them through headphones? That's because they are coming directly into our ears without involving the outer ear – in particular the upper part, which is known as the **pinna**. The shape of the pinna directs sound inwards, and that helps us to judge where the sound is coming from. Everyone's ears are slightly different, so there are small differences in how the sound is channelled inwards, and our brains are sensitive to these. Researchers investigating those differences have made models of the shape of the pinna and recorded sounds using microphones placed in those models. They also recorded other sounds without a pinna shape, and played both types of sound back to their research participants (Wenzel et al., 1993). People were much more accurate at judging where the sound was coming from when hearing the sounds recorded as if they had been received by a human ear. And they were most accurate of all when judging sounds recorded from a model made from their own ears.

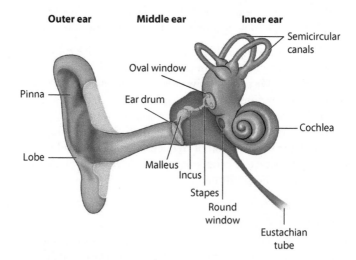

Figure 4.2 The structure of the ear

WHAT HAPPENS IN THE EAR?

The vibrations the ear receives are very small, so they need to be amplified before they are converted into signals the brain can use. We do this using a set of small bones in the middle ear, known as the malleus, incus and stapes (Latin for hammer, anvil and stirrup). Figure 4.2 shows how the ear is structured. Sound waves are collected by the pinna and focused on the eardrum, or tympanic membrane. These sound waves make the tympanic membrane vibrate, like a drum. The vibrations are picked up by the malleus bone, which amplifies them a bit, then passed on to the incus which amplifies them a little more, and then to the stapes, which does the same. The stapes is in contact with another membrane, known as the oval window, and by this time the vibrations have become larger and more detectable. The oval window passes them on to the inner ear, which is filled with a fluid called perilymph. Fluid doesn't squash as easily as air, so there's another membrane lower down, called the round window, which moves in and out to compensate for the pressure caused by the movement of the oval window.

The purpose of the ears, then, is to collect sound waves and amplify them. But the brain, as we know, works on electrical impulses, not pressure, so the next step is **transduction**: converting the sound waves into electrical impulses that can be passed on to the brain. This is what happens in the cochlea of the inner ear. Figure 4.3 shows how the cochlea is divided into sections. The vibrations from the oval window 'push' the fluid, causing the membranes of the inner ear to vibrate in their turn. A series of hair cells on one of the membranes, the basilar membrane, respond to the vibrations by generating an electrical impulse.

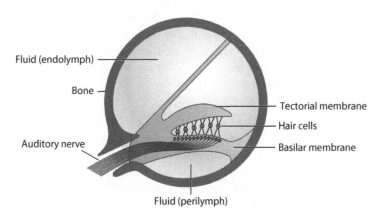

Figure 4.3 The structure of the cochlea

Not all the hair cells respond in the same way. The hair cells at one end of the basilar membrane are more sensitive to higher frequencies of vibrations – what we experience as higher-pitched sounds – while the other end is more sensitive to lower frequencies, and the ones in between respond to middle-range frequencies. This means that the sound we are receiving will stimulate different hair cells to produce electrical impulses, depending on what the sound is like. The hair cells make connections with a set of neurones with very long axons. These bunch together and form the auditory nerve, and the auditory nerve passes the information – now in the form of electrical signals – to the brain.

WHAT HAPPENS IN THE BRAIN?

As with vision, much of the incoming information is sorted before the signals actually reach the auditory cortex – the part of the brain that makes sense of what we are hearing. The auditory nerve has its first synapse point in the thalamus, at a group of cells known as the medial geniculate nucleus. From there, the information goes on to a group of cells found at the side of the brainstem known as the cochlear nucleus. This is where the sound frequencies are identified. The upper parts of the cochlear nucleus receive information about high-frequency sounds, while the lower parts receive information about lower frequencies, and sounds of intermediate pitch stimulate cells in between.

We find some sounds relaxing, and as a rule these tend to be the middle-range sounds. Other sounds, though, we find quite disturbing. We react more strongly to a scream – a high-pitched sound – than we do to a mid-range hum. And we also react to some kinds of low-frequency sounds; think of the 'shark' theme in the *Jaws* films, for example, or the sound of a lion's roar. So we can see that our ability to distinguish high-pitched sounds from lower-pitched ones could have real survival value, which may be why it is one of the first things the brain processes.

Being alerted to potential danger is all very well, but we also need to know where the sound is coming from. So some of the nerve fibres coming from the cochlear nucleus go to the pons, which helps us to work out the direction of a sound's source. They go to a part of the pons called the **trapezoid body**, which forms a crossover point for the auditory nerve fibres, a bit like the optic chiasma does for our vision. Like the optic chiasma, half the nerve fibres carrying sound information cross over to the other hemisphere, and half stay on the same side. This means that the brain can compare the slight differences between the sounds coming from each ear, which helps us to work out where a sound is coming from.

The next stop in the auditory pathway is also in the pons, at a part called the superior olivary complex. Here, the information is sorted even more, beginning to give us some of the richness we experience.

One side of the superior olivary complex processes loudness, with some cells responding to high-amplitude vibrations while others respond to quieter sounds. The other side of the superior olivary complex is all about timing, with some cells responding to rapid repetitions of sounds and others to single notes or calls. Once it has been sorted for these basic characteristics, auditory information passes to the cerebrum, to the auditory cortex on the side of the temporal lobe. That is where our conscious hearing is processed (see Figure 4.4). These aren't the only connections: sound information makes many other connections with other parts of the brain as well, but what I have described are the main pathways.

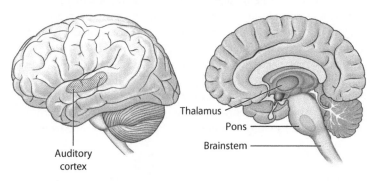

Figure 4.4 Auditory processing areas of the brain

Making sounds make sense

Because of the route that the information has taken, the signals that the auditory cortex receives include information about pitch, loudness, timing and the direction that a sound is coming from. All of these contribute to our experience of sounds. The task of the auditory cortex, and the other brain areas associated with it, is to make those sounds make sense. Pitch, as we've seen, is fundamental to how we interpret sounds, and in the primary auditory cortex the central area processes low-pitched sounds while the outer areas process the higher pitches. This makes what researchers call a **tonotopic map** – that is, an area that responds to different pitches in different places. But the rest of the information shapes what sense we make from the whole sound.

The auditory cortex (see Figure 4.5) is generally organized into three areas: the core, the belt and the parabelt. The core area is the primary auditory cortex, which responds to specific features of the sounds. It contains specialized neurones, which collect from neurones responding to single frequencies and in their turn respond to mixtures of frequencies. These in turn feed information to other neurones. Some cells, for example, respond to voices more actively than they do to pure sounds. Other cells respond only to changes in frequency but not to the frequencies themselves. And some neurones respond by becoming active in response to specific frequencies but switching off and becoming inactive when stimulated by a similar but not identical one. It's all a bit like the simple, complex and hypercomplex cells of the visual cortex, but the response is to the qualities of complex sounds rather than visual shapes.

The belt area, as its name suggests, goes round the middle. It is the part immediately surrounding the core, and it seems to do two things. The front part codes the content of the sound and passes that information on to the frontal and parietal lobes of the cerebrum. The back part of the belt region, known as the planum temporale, works out where the sound is coming from. It collates the information from the trapezoid body with the information coming from the distinctive shape of the pinna. So neural processing in the belt area helps to tell us what we are listening to and where the sound is coming from.

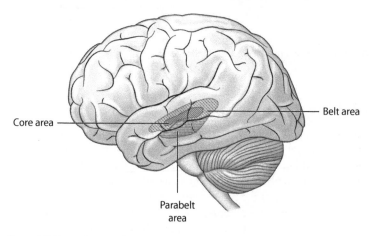

Core area

Belt area

Parabelt
area

Figure 4.5 The auditory cortex

The belt area of the auditory cortex merges into the outer area, the parabelt, which deals with more complex functions of hearing. One of the things it helps to do, for example, is identify and interpret speech, which we'll look at next. But it has other functions, too: it's concerned with auditory working memory, which is the way you carry the immediate memory of a sound in your head while you are thinking about it; with linking together auditory and visual information; and with making connections with our memories of other sounds.

If the primary auditory cortex is damaged on just one side of the brain, we experience some hearing impairment, so we don't hear so well and have difficulty locating sounds. If it is damaged on both sides, though, we experience full deafness. That deafness is not the result of sounds being undetected – it arises from the fact that the brain can't make sense of them. As we've seen, though, sound information has already been processed to an extent before it reaches the cortex, so people with this type of deafness can react emotionally or reflexively to sounds. They might show startle reactions to a sudden loud sound, for example, or be emotionally disturbed by loud cries. What they can't do is describe in words what it is they are reacting to; as with blindsight, they experience the feeling but not the perception of the stimulus.

The auditory cortex has many connections, but it links with other parts of the brain through two main routes. One of them, known as the 'what' route, is all about identifying specific sounds. It is the way we work out what we are actually hearing. It's sometimes known as the **ventral stream**, and it passes along the lower side of the auditory cortex and the front (anterior) part of the temporal lobe. The other is known as the 'where' route and, as its name suggests, it's about locating where sounds are coming from. It passes along the upper, or dorsal, side of the auditory cortex, so it's often called the **dorsal stream**. This neural pathway makes connections with motor parts of the parietal lobe and the frontal cortex – preparing us for movement, among other things. Both of these routes, of course, are equally important in helping an individual to survive. But there's more to hearing than this, so let's look at some of the more specific things that we listen to.

Hearing speech

Since the nineteenth century, scientists have been aware that the left hemisphere of the brain is more important for language and speech than the right hemisphere. It isn't totally exclusive: the right hemisphere does have some simple language functions and, as we saw in Chapter 2, some studies of people who have had their whole left hemisphere removed have shown surprising recovery of language functions – not just conversational speech but also memories of songs and stories. So it isn't as simple as left brain = language, right brain = music, but then it never was. We'll be looking at how we process and use language in more detail in Chapter 10.

Have you noticed how easy it is to tell the sound of voices from other sounds? We can distinguish speech from other types of sound even when we can't make out the words. Part of this has to do with learning: we become familiar with the speech sounds (phonemes) used in our own language, possibly even before birth. Different languages use different phonemes, and we may not always recognize those particular sounds as part of a particular language. But we can still recognize speech when we hear it, even with unfamiliar phonemes, because we are also familiar with the typical timings and patterns that people use when speaking. Anyone who has known a small child while they are learning to speak will have seen how they often get the patterns and timing of speech exactly right, even while the 'words' they are producing are still only babble or nonsense sounds.

There comes a point in our processing of sounds where we differentiate between speech and other types of noise. Functional imaging studies show us that both sides of the primary auditory cortex respond equally to speech noise and other types of noise. But then the 'what' pathway begins to distinguish between the two. Speech sounds produce more neural activity in the left temporal lobe than they do in the right temporal lobe, while the right temporal lobe responds more strongly to changes in pitch, particularly musical ones. There is activation in the opposite sides for both, but it isn't as strong.

The 'what' pathway branches again as it leaves the primary auditory cortex. One branch passes through the anterior temporal lobe, connecting with a speech recognition area below the primary auditory cortex and going on towards the frontal lobes. On the way, it passes through an area that deals with knowledge of words and meanings and another area that is involved with the planning of speech. The other branch goes towards the back of the temporal lobes and makes connections with areas which also link with the visual system – with an area known to interpret gestures and another concerned with reading. So while both branches are concerned with what the signals are (as opposed to where they are coming from), the first branch is concerned with the mechanics of how speech happens while the second is concerned with meaning. Together, they allow us to understand what people are saying when we listen to them talking and watch them as they do so.

Hearing music

Making sense of what we hear has everything to do with the timing of sounds as well as the sounds themselves. This is particularly apparent when we consider how we respond to music. Early advertisements for the gramophone, dating from the first half of the twentieth century, described it as 'The Musical Instrument Everyone

Can Play'. In modern life we are surrounded by music in just about every sphere, and it's easy for us to forget that this is a relatively recent happening. Until the advent of popular and affordable recorded music, most people's exposure to music was very limited. Unless they were lucky enough to have an instrument at home and the skill to be able to play it, most people's experience of music was only what they heard at fairs, concerts and other social events.

Yet we have always responded strongly to music, and it has always been an important part of human society. Weary and wanting to recover from her domestic cares, Meg Brooke in the novel *Little Women* (1869) asks her husband if they can go to a concert, because 'I really need some music to put me in tune', showing how music can relax, soothe and calm us down. It can also have the opposite effect, of course, as composers of film music are well aware, and we will be looking again at the emotional effects of music in Chapter 8. But we all appreciate music naturally. Even small children enjoy hearing music, and some researchers believe that music perception is as strongly built into the human brain as the ability to use language.

We can think of music as being patterns in time. Our vision is sensitive to patterns in light. Our hearing is also sensitive to patterns but, since sound is ephemeral, we perceive the patterns in how notes combine or follow one another. As we've seen, the auditory system is far more sensitive to timing than the visual system, and timing is an essential part of music. Judging by the effects of lesions in particular areas, the right hemisphere is more concerned with processing information about pitch, while the left hemisphere processes information about timing. Unlike the popular myth, then, both cerebral hemispheres are involved in music perception, but they generally focus on different aspects of the phenomenon. In fact, it is generally considered that music has its overall positive effect because it stimulates areas across the whole brain, rather than just one or two parts of it.

You'll have gathered from this that music processing is quite a complex affair. Brain-imaging studies of people listening to music show activity in the auditory, motor and limbic systems. At its simplest level, the auditory system is processing melody, the motor system is reflecting timing (although we know that timing is also processed in the auditory system), and the limbic system is responding to the emotional nuances of the music. Incidentally, these imaging studies found similar effects regardless of whether people were listening to pop or classical music. Music can also help our general brain functioning: studies of older people found that they showed faster neural processing and a better memory for events as a result of listening to music.

Case study: Temporal lobe damage

It has been shown that we process music in the temporal lobes of the brain. One woman who had suffered damage to both temporal lobes found herself unable to identify tunes, even the ones that had been most familiar to her before her accident. She could distinguish tones of voice, showing that she could still react to sounds of different pitch, and she could recognize voices and other sounds in the general environment. But when it came to tunes, although she could enjoy them while they were happening, she simply couldn't recognize them. It appeared that it was her memory for music, rather than the ability to hear it, which was impaired, although the rest of her memory seemed unaffected.

Like language, music can take different forms in different cultures, but it always tends to be based on a set of discrete pitch levels – what we think of as a musical 'notes', clearly separated. The range of possible wavelengths in between is regarded as 'off true' and not appropriate for music. While musical pitches are generally arranged in octaves, what counts as an octave can vary from one musical scale to another. Western music, for example, usually uses a heptatonic scale, with seven notes in an octave. But there is no single universal scale: older Celtic and Gaelic tunes often use a pentatonic scale based on five notes, for instance, as does much Asian music; and some supposedly 'primitive' (but actually very sophisticated) music is sometimes based on scales of just four or even three notes.

These scales are not random. The way the notes are grouped is based on the properties of their sound, as processed in the auditory system. Some combinations 'feel' right, some notes seem to follow others 'naturally', and those impressions come from a combination of the acoustic properties of the sound and the cultural experiences of the individual. It is not that we don't perceive the frequencies between the notes: as we've seen, speech perception involves subtle nuances of pitch, not limited to the tones which are parts of musical scales. It's that they don't seem to 'fit' properly, so they sound wrong in the context of the music.

Amusia is the inability to process music. It can happen as a result of brain injury or disease, but some people have congenital amusia – that is, they are 'tone deaf' and unable to sing accurately or produce music themselves. They can still distinguish frequencies, though, because they are still able to copy pitch shifts in speech. People with amusia are more sensitive to speech tones than they are to music. Being tone deaf is associated with the density of grey matter in the right hemisphere around the auditory cortex, but it doesn't correlate with any other known brain deficit. However, brain plasticity studies show us that the amount of grey matter can reflect our experiences, so the fact that they have less grey matter in these areas could have resulted from the fact that those people simply don't engage with music as much. In other words, it could be a result, rather than a cause, of the amusia. We will learn more about it as research in this area continues.

MUSIC AND DANCE

Most people have some experience of performing music, even if it is only singing along to a favourite song in the car or joining in a football chant. Singing, of course, combines our knowledge and processing of speech with our knowledge of music, and even the simplest level of singing requires the brain to exercise control of timing and pitch. Even tone-deaf people can respond to musical timing: it's quite usual to tap one's foot to a beat or clap along with some kinds of music, and this reflects the strong connections between auditory and motor processing. The fact that we don't respond in the same way to rhythmical visual input (like, say, a video of a ball bouncing) shows us how strong that connection is.

Figure 4.6 Dance happens in all cultures

It also shows us why dance is such a powerful and universal part of the experience of being human. The close links between the movement centres and the auditory processing areas of the brain give us a powerful tendency to respond to rhythmical sounds with movement. It's inbuilt and universal: all human cultures, no matter how remote, have some tradition of dance and music. Dance connects our auditory input with muscle actions, and also with input from the part of the inner ear that gives us our sense of balance (see Chapter 5). Moving to music comes naturally to us, but some forms of dance are more complex, of course. As with other physical skills, as we practise we become better at it, and that practice develops neural connections in the brain. Studies have shown that the combination of mental and physical exercise involved in dancing can sharpen our problem-solving abilities and working memory, as well as helping us to keep fit.

Whether it is dancing or playing a musical instrument, performing is doing. We'll be exploring the connections between motor activity and musical performance in Chapter 6, which is all about movement. But when we think of musical performance we are generally talking about something much more sophisticated than just tapping our feet to a tune. Playing an instrument like a piano requires the brain to connect a spatial activity – choosing which specific keys to press – with an auditory stimulus, the note, in a way that is precise, unthinking and rapid. This means that a pianist has to forge more or less automatic connections between the motor cortex, the cerebellum, the auditory cortex and also the frontal cortex, which is involved with planned actions and decision-making. Brain scans of professional musicians show that their brains have changed in response to their years of continuous learning. They have larger areas in the motor cortex as a result of their extensive physical practice, and larger and more symmetrical auditory areas.

Trained musicians also process music differently in the brain. For ordinary people, listening to music mainly (but not entirely) involves the right hemisphere of the cerebrum. When trained musicians listen to music, though, both hemispheres are active, and the left hemisphere is often more active then the right. Professional musicians also have a more developed corpus callosum, the band of fibres connecting the two hemispheres and passing information from one to the other. We know that the left hemisphere is more concerned with computational activities, and it may be that musicians bring more formal knowledge to their listening than an ordinary person, because they are familiar with musical syntax. While a trained musician will be unconsciously analysing the music they are listening to, a non-musician will simply be experiencing the music without analysing it.

Timing, as we've seen, is an essential part of music, and it is one of the things that musical training focuses on. People who are musically trained are particularly sensitive to timing, and have been shown to process all sorts of auditory information more quickly as a result. Very quick speech, for example, requires rapid neural responses in order to decode it, and so does other fast auditory input. Older adults quite often find difficulty in the neural processing of this type of input as their brain cells respond more slowly; but trained musicians don't suffer from this problem. Nor, it seems, do people who had some musical training when they were children. A study by White-Schwoch et al. (2013) compared older adults who had learned an instrument in childhood with others who had not had any musical training. Even those who had given up playing as they got older showed faster neural responses to speech than those who had never learned. It seems that the benefits of music education can stay with you, even 40 years later.

Hearing, then, is a complex sense that is an important part of our experience of being human. Together with sight, hearing allows us to detect, identify and respond to the world around us.

Focus points

✻ Sound information is collected, amplified and transduced into electrical impulses by the ears. It then passes to the thalamus and the auditory cortex.

✻ The centre of the auditory cortex codes the information, while the belt area works out what and where it is, and the parabelt connects it with memories and other senses.

✻ We are particularly primed to detect speech, and can easily recognize the sounds of voices above other sounds.

✻ We are also powerfully primed to respond to music. Each culture has its own musical traditions, but our hearing system learns them easily. Both sides of the brain process music, but in different ways.

✻ Hearing and movement are strongly linked, so we naturally respond to musical rhythm by movement and dance. This is why dance is found in all human cultures.

Next step

As we have seen, sight and hearing are the two most important senses for humans, but we have several others as well, and we will look at these in the next chapter.

5

Experiencing other senses

In this chapter you will learn:

► *all about our sense of smell and how it works*

► *how our sense of taste makes us sensitive to a range of tastes and textures*

► *how the sense of touch detects pressure and temperature*

► *how we feel pain and how we can learn to control it*

► *that there are many kinds of sensory illusion, including phantom pain and synaesthesia.*

Have you ever smelled something that brought back a flood of memories? Or touched something which felt quite different from the way it looked? Vision and hearing are our most important senses, but we have several others as well (see Figure 5.1). The ancient Greeks identified three more – touch, taste and smell – so people often assume that we have five senses. In reality, though, we have many more. We have different sense receptors for heat, for movement, for pain and for balance, for example, and we know something (although not everything) about how the brain processes each of these. In this chapter we will look at our other senses of smell, taste, touch, and pain. Then we'll go on to look at sensory illusions such as phantom pain, and what we know about synaesthesia – when our senses overlap or seem to replace one another.

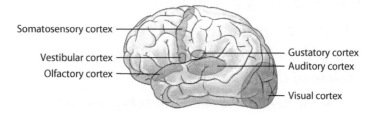

Figure 5.1 Sensory areas of the brain

The sense of smell

We are bathed in a sea of information, all the time. Our vision detects a small proportion of the electromagnetic radiation that surrounds us; our sense of hearing detects certain wavelengths of the vibrations in the air around us; and our chemosensory system – the senses of smell and taste – detects chemicals, in the air around us and in the things we put into our mouths.

It is likely that the sense of smell was the second of the main senses to evolve. Taste was probably first, of course, because it's an immediate reaction to what is in contact with us. Smell, though, allows us to locate things at a distance. For the earliest animals, smell and taste were probably the same: an animal swimming in the primeval oceans would sample the chemicals in the water around it, and use that information to detect aspects of its surroundings. Even a simple one-celled animal like an amoeba makes slight chemical changes to the water it swims in, so having a sense that can detect chemical traces would be an important survival trait – both for detecting food and for avoiding becoming it.

For many animals, smell is their most important sense, as is confirmed by its location within the brain and the size of their **olfactory bulbs** – the parts of the brain used to interpret smells. The olfactory bulbs are part of the limbic system of the brain rather than the cerebrum, which also implies that they have an older evolutionary origin. In some animals, like fish, the olfactory bulbs are so enlarged by comparison with other parts of the brain that they stand out as distinct organs (see Figure 5.2). In humans, they are simply part of the general limbic system, underneath the cerebrum, doing some processing and making links with other areas of the brain such as the amygdala and the olfactory cortex.

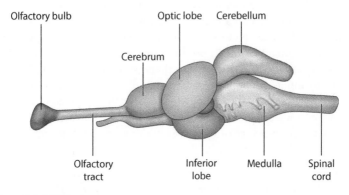

Figure 5.2 A fish brain

HOW DOES THE SENSE OF SMELL WORK?
The sense of smell begins in the olfactory epithelium, which is a series of layers of cells lining the nasal cavity, and covered with mucus. The mucus dissolves chemical molecules in the air entering the nose, and these chemicals are detected by neurones embedded in the epithelium. These neurones have small hair cells, known as cilia, which are sensitive to particular chemicals. They produce an electrical response when they receive them, and pass that impulse on to the olfactory bulbs.

In humans, the olfactory bulbs are the first stage of processing smells. We don't have anything like the same depth of understanding of how the brain codes smells as we do with vision, but it is thought that the olfactory bulbs have four general functions. One of these is to enhance the sensitivity of detecting smells, for example by firing more rapidly when they identify stronger ones. Another is to distinguish between different smells, identifying their different components and classifying them accordingly. A third function is to act as a kind of filter, emphasizing newer or more pungent smells and filtering out background ones. And the fourth is to make connections with our brain mechanisms of alertness and arousal, which helps to emphasize smells that might signal something potentially harmful.

It is for this reason that smells can be so powerful in the emotions. The olfactory bulbs have direct connections to the amygdala, which is the centre of our emotional responses in the brain, and also to the hippocampus, which is particularly concerned with memory, as well as to the olfactory cortex and other brain regions. Its direct connection to the hippocampus shows us how smells can be so powerfully associated with special memories, and its connections with the brainstem and amygdala show how smell can have such a powerful influence on our emotional responses. Aromatherapy is the use of the aromas, or smells produced by essential oils to influence the body, such as by encouraging calmness or alertness, and it works by influencing these extensive and direct connections of the olfactory system. It's important to remember that these connections are direct: they don't come through the olfactory cortex, so we may not even be aware of what we are experiencing. But smells can affect us in very subtle ways.

The olfactory bulbs, then, have connections to many other parts of the brain, and one of them is the olfactory cortex. This is an area of the cerebrum that lies along the toe of the temporal lobe and links with other areas of the cerebral cortex. It also contains an area known as the **olfactory tubercle**, which is one of the best-connected areas of the brain, with more than 20 different sources of incoming information. The olfactory tubercle receives input from the olfactory bulbs, the thalamus, the amygdala, the hypothalamus, the hippocampus, the brainstem, the retina and the auditory cortex, and many other areas of the brain. It also has a similar number of pathways sending information outwards across the brain (see Figure 5.3). All this, of course, tells us why our sense of smell can give us such a complex range of experiences. The olfactory tubercle is directly associated with feelings of reward and pleasure, with arousal (both alarm-related arousal and sexual arousal), with attention, and with many other aspects of our experience.

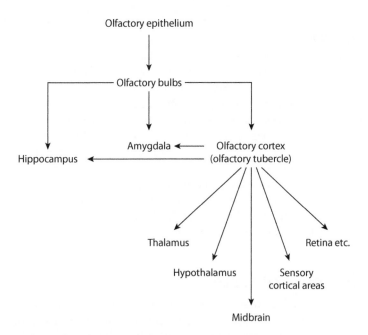

Figure 5.3 Smell pathways in the brain

Some people have an extremely acute sense of smell. It is just being discovered, for instance, that some people are even able to detect disease in other people through the changes it makes to the smell of their bodies. Other people, by contrast, have no sense of smell at all, or almost none. This is known as **anosmia**, and we know much less about it than we do about blindness or deafness. Congenital anosmia, where someone is born with no sense of smell, though, is quite rare. Most anosmia results from damage of some kind, either to the relevant brain areas or to the nasal epithelium. And, of course, we can experience temporary anosmia as a result of catarrh or colds. But even if we are not aware of it, our sense of smell may still be active in the brain: an anosmic person may experience a sense that something is wrong even if they can't identify the smell itself – like blindsight or the ability of deaf people to react to sudden noises. The extensive connections of the olfactory system, both from the olfactory bulbs and from the olfactory cortex, show us just how powerful the sense of smell is and how important it has been in our evolutionary heritage.

Key idea

Humans who have been brought up in direct contact with their natural world usually have a very keen sense of smell. The traditional upbringing of Native Australian children, for example, involves explicit training in smell detection and recognition, in much the same way as children in industrialized environments are trained in hearing and vision through rhymes, colours and pictures. English as a language is a product of an industrialized world, so we don't have all the words we'd need to describe what this kind of smell training might include. Knowing what we do about brain plasticity and how the brain develops with training, though, it is likely that the olfactory cortex, which interprets smells, is better developed in Native Australians who have experienced a traditional upbringing than it is for a typical European, and at least it would have more neural connections.

The sense of taste

You may often hear that the sense of taste and smell are the same – that if your nose is blocked, you can't taste anything either. However, the sense of taste, or **gustation**, is different from the sense of smell. It has different receptors and is processed by a different part of the brain. The two are often linked, though, because there are some smell-detecting cells at the back of the throat, which combine their information with that produced by the taste buds in the tongue. Some people feel that losing their sense of smell also means losing their sense of taste, but what it really means is that it has become less intense, because they are relying on their gustatory sense alone where they've been used to augmenting it with smell. People with less reliance on the sense of smell, such as those who have experienced chronic catarrh or regular nasal infections while young, find their sense of taste less impeded when they have a cold because they are more used to concentrating on the gustatory sense alone.

Our sense of taste begins with taste receptors, known as taste buds. There are thousands of these in the mouth: most of them are located in small bumps on the tongue, but others are all over the back, sides and roof of the mouth, and also in the throat. Opinions differ as to how many types of taste receptors there are: the five basic ones of salt, sweet, sour, bitter and umami (savoury) are well established, but some researchers now believe that there is a sixth type of receptor which responds to fattiness. The various combinations of receptors means that we can detect a very wide range of gustatory experiences, both pleasant and unpleasant.

In terms of our human experience, of course, taste is also influenced by many other factors. We've already seen how the smell of something can influence its taste, but so can its temperature, texture, pungency (the 'hotness' of chillies or mustard) and 'coolness' like that produced by mint. Put all that together with social factors like eating with friends, or the special ambience of a nice restaurant or special meal at home, and we can see how our gustatory sense can be such a powerful motivator in our lives.

Taste buds contain specialized cells, which react to different combinations of chemicals by producing an electrical impulse and passing it along an afferent nerve to the medulla. This direct connection with the medulla shows how very basic our sense of taste is: as we saw earlier, it was probably the very first of our 'external' senses; and we saw in Chapter 1 how the medulla was one of the earliest parts of the brain to evolve. From the medulla, information about taste passes to an area at the back of the thalamus, which in turn connects with the primary gustatory cortex.

Studies of the gustatory cortex tend to define it as having two parts: the frontal operculum, at the base of the frontal lobes of the brain, and the anterior insula, which is on the folded-under part of the frontal lobe. However, these two parts are continuous and appear to work together, rather than being separate. Neurons in this area have been shown to respond to salt, sweet, sour and bitter stimuli, and also to indicate by their rate of firing just how strong those particular tastes are. Studies of how taste neurones respond to stimuli have also shown that they are linked with neurones that indicate feelings of pleasantness, both in taste and with other types of reward. All of this implies that how we react to taste, in terms of whether we find it pleasant or not, is a fundamental part of the sense rather than just being a later add-on.

This links with the evolutionary purpose of the sense of taste, of course, which is to alert us to what might be good or bad to eat. Sweetness usually indicates energy-rich food, while bitterness indicates potential poisons or toxic substances that are best avoided. In nature, foodstuffs that are both fatty and sweet are quite uncommon, but they would have been extremely useful sources of nutrition. It has been suggested that this is why we find food that combines sweetness and fats so attractive – that is, cakes, puddings and sugary snacks. In our modern world food is more abundant, so we don't need to home in on these types of sources of energy; but some researchers believe that we have retained a taste for them from our evolutionary past.

Taste is shaped by learning and culture, as we know. Part of the joy of living in a multicultural world is our exposure to a variety of different tastes from different cultural groups. Some cultures emphasize highly pungent foods, for example, while others focus on sweetness or complexity of taste. But in all cultures we tend to avoid tastes containing noxious compounds. This is, of course, the other survival value of gustation, which is to make sure that we avoid things that are harmful or poisonous. While there is cultural variation in how people appreciate different tastes, just about everyone finds bitter tastes unpleasant. This is partly because those tastes usually indicate the presence of nitrogenous compounds, which could damage the body if consumed in quantity. Unripe fruits or berries and certain plants can also be toxic, and they usually taste unpleasant. So our tendency to avoid nasty tastes has been fundamental to our survival as a species.

The sense of touch

The sense of touch is located in the skin, and it is included in the group of senses known as **somatosensation** ('somato' means 'to do with the body'). Touch is actually one of three different senses in somatosensation – or, at least, there are three different types of receptors, which respond to three different types of stimulus. One of these is **mechanoreception,** which is how we discriminate between different types of touch, like a light stroking or sustained pressure. That's the main sense that we tend to think of as the sense of touch. But we also have **thermoreception**, which is the way we perceive heat and temperature; and a third source of information from receptors in the skin is **nociception,** which is the way we perceive pain. We'll look at nociception later in this chapter.

As you'll know from your own experience, we can be touched in several different ways, and our skin's receptors transduce these into electrical impulses. Our skin has several different layers, and light touch, steady pressure and texture are all detected by receptors relatively near the surface, in the hair root and the upper layer of skin, known as the epidermis. Some of these receptors are free nerve endings in the hair root, while others are specialized receptors with disc-like endings in the epidermis. Different receptors just below the epidermis detect sensations like fluttering or stroking, while receptors in the next layer down, the dermis, respond to touch involving vibration or stretching (see Figure 5.4).

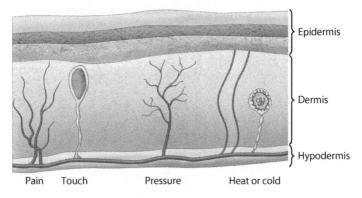

Epidermis

Dermis

Hypodermis

Pain   Touch   Pressure   Heat or cold

Figure 5.4 Sense receptors in the skin

Thermoreception – the sense of temperature – comes from the stimulation of free nerve endings in the skin, which respond extremely rapidly to changes in temperature and pass that information directly to the spinal cord. The spinal cord can then initiate an avoidance reflex, if necessary, contracting our muscles so that we pull away from something too hot or too cold. Other touch receptors pass their messages to the thalamus (via the spinal cord), which makes us aware if something is too hot or too cold. From the thalamus, the information is passed to the somatosensory cortex. This is a strip of the cerebral cortex running along the very front of the parietal lobe, opposite a similar strip – the motor cortex – running along the back of the frontal lobe. Different parts of the somatosensory cortex respond to messages coming from different parts of the body, and more sensitive areas have more of the cortex devoted to them.

We have quite a detailed knowledge of how the somatosensory cortex of the brain responds to tactile sensations. It's arranged in a sort of body order, with the toe at the top of the sensory strip of cortex, and the mouth at the bottom. The first mapping of this area came from stimulating the brains of people undergoing open-brain surgery, who would report feeling as if a particular part of the body had been touched when a particular area was stimulated. The 'sensory homunculus' shown in Figure 5.5 was a summary of the findings of a pioneer in this area, the neurosurgeon Wilder Penfield, and it illustrates the sensitivity of different parts of the body. It is distorted and unlike the usual body because it reflects the levels of sensitivity of different areas, with larger parts being more sensitive. The torso, for example, is quite a large part of the body, but it takes up less of the somatosensory area than the lips because it is so much less sensitive. Our fingers and hands are very sensitive, too, so they are larger in the homunculus, taking up quite a lot of space on the somatosensory cortex – much more, for instance, than the upper arms.

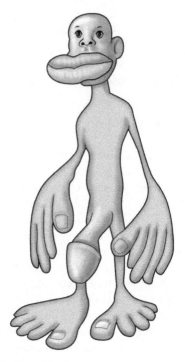

Figure 5.5 A sensory homunculus

Feeling pain

Our sense of touch can be closely linked with our feelings of pain. Pain receptors are mostly located in the skin, although pain is also part of the sense known as **interoception**, which has to do with internal sensations like stomach pain and the movement of the internal organs. We do have pain receptor neurones in our internal organs, joints and muscles, but not as many as we have in our skin. This is why a feeling of internal pain can be deceptive if we're trying to work out where it is actually coming from.

Pain receptors are called **nociceptors**, and they respond to three kinds of stimulation: mechanical, like cutting or crushing; thermal, like intense heat or intense cold; and chemical, like the sting of mustard powder on the skin or a trace of chilli powder in the eyes. The nerve endings of nociceptor neurones respond to these painful stimuli if they are strong enough, and pass the information to the spinal cord. This produces a direct pain reflex, sending messages to the muscles to contract, and also passes the information on to the brain. You don't have to think to pull your hand away from something that stings – your nervous system does it automatically. As you can see, this type of reflex is a useful survival mechanism for minimizing damage.

Stimulation of nociceptors in the face passes the information to the trigeminal nerve, which acts like a smaller 'spinal cord' for the head. But most nociceptors pass information to the spinal cord itself, where the pain information is separated into several streams. Some neurones carry information from the spinal cord to the thalamus, some take it to the brainstem, and some take it to the medulla, pons and the periaqueductal grey matter – an area within the midbrain. This is because there are different pathways for different types of pain. For example, while both types of pain pass through the brainstem, sudden, short-term pain takes a lateral route, going along one side of the brainstem, while chronic pain like long-term backache takes a medial route, going directly through its central core.

Brain scans of people experiencing pain have mostly, although not exclusively, tended to focus on short-term painful stimulation. Researchers have identified what has become known as the **pain matrix** – a collection of areas in the nervous system which are all activated when people are in pain, and which all seem to work together. The thalamus is an active part of the pain matrix, and it also passes information on to a part of the cerebrum known as the insular cortex. The insula, as we saw in Chapter 1, is folded deep within the groove that separates the temporal lobe from the frontal and parietal lobes. It is known to be involved in perception and motor control, as well as many other functions, which might explain why pain sometimes makes us want to move about. Other neurones transmit pain information from the thalamus to the cingulate cortex, which surrounds the front part of the corpus callosum.

We can see, then, that there is no single area for pain on the cerebrum. Instead, it involves several brain areas, but particularly the insula and the cingulate cortex (see Figure 5.6). This is interesting because, apart from its role in perception and motor control, the insular cortex is actively involved in consciousness, and also plays a part in our experience of emotions. Its role in the perception of pain is believed to be to separate pain information from less significant feelings like itching or heat, letting us know that it is more serious than that. Activity in the cingulate cortex, on the other hand, is thought to be why we feel pain to be unpleasant. Researchers can't be absolutely definite about these functions, though; partly because there are so many different kinds of pain, partly because the brain itself is so complex in its functioning, and partly because everybody is different in the ways that they think about pain.

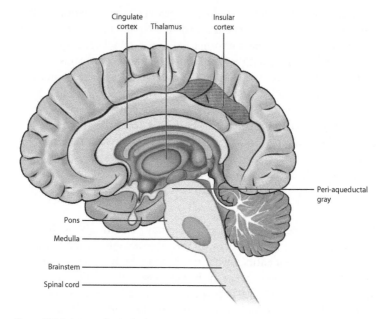

Figure 5.6 Pain areas in the brain

Another way in which the anterior cingulate cortex regulates our feelings of pain is through its connections with the grey matter in the midbrain, which is particularly rich in natural opiates known as endorphins and enkephalins. These are neurotransmitters which act as natural painkillers, and are released at times of extreme shock, high physical stress or vigorous exercise. Powerful opiate drugs like heroin and morphine are effective because they imitate these natural painkillers in the brain (more about that in Chapter 13), although their side effects mean that they are not always an effective way of tackling pain. They tend to be more commonly used to alleviate the pain of terminal cancer, since problems of addiction are distinctly less important for those people than relieving the pain they are experiencing.

We often refer to social experiences as having been 'painful'. An embarrassing episode, the sense of loss we experience from a serious quarrel with a loved one, loneliness and social rejection are all experiences we would describe as painful. But this seems to be more than just a metaphor: some interesting evidence has shown that social pain of this kind activates the same areas of the brain as are activated by physical pain: the insula, the cingulate cortex and other areas as well. So perhaps the poetic descriptions of experiences like bereavement or social rejection as 'painful' may be more accurate than we think. We'll come back to this idea in Chapter 9.

CONTROLLING PAIN

Some people with chronic pain have been given electrical implants in the periaqueductal grey matter, which they can activate to stimulate the area directly. For some, this gives them a degree of control over their pain. But not all chronic pain is controllable in this way. In 1982 Melzack and Wall proposed that the brain has its own 'gates', which may or may not allow pain information through, depending on past or current experience. A gate can be closed by neural impulses coming down from other parts of the brain, like the impulses produced during extreme arousal or interest. Other kinds of stimulus, like pressure, can sometimes close the gate as well, which might explain why rubbing a painful spot can sometimes help to relieve pain.

Some researchers have investigated whether people can learn to control pain using mental strategies. This is a strategy often used by experienced athletes, who need to drive their bodies far beyond their comfort zone in order to achieve peak performance. A meta-analysis of 40 different studies by Fernandez and Turk (1989) showed that developing mental imagery techniques can have a definite effect on controlling pain. More recently, Woo et al. (2015) explored people's brain activity in response to painful heat stimuli on their arms. They had three conditions to the study: in one condition the participants were asked to clear their minds and not think of anything; in the second they were asked to imagine that the burning heat was actually damaging their skin (it was uncomfortable, but not damaging); and in the third condition the participants were asked to imagine that the heat was a pleasant experience on a very cold day.

In all three conditions brain scans showed the same activity in the physical pain pathways. The three conditions also produced the expected reactions in the participants: their perception of pain increased in the second condition and decreased in the third. But the conditions also activated a second brain pathway, depending on how the participants were thinking. This pathway involved the nucleus accumbens and the ventromedial prefrontal cortex – areas of the frontal lobes of the brain already known to be involved in motivation, valuation and emotional appraisal, but that hadn't previously been associated with pain control. This, and supporting research, showed, as researchers and athletes had long suspected, how cognitive strategies like imagery, distraction and maybe even hypnosis can affect the brain's activity in ways that can directly influence the pain we are experiencing.

The general model we have of how pain affects us is that it has three different dimensions. The first of these is the sensory-discriminative dimension, which is all about how strong the pain is, where it is, and how long it lasts. The second is the affective-motivational dimension, which has to do with the feeling of unpleasantness and our urge to escape from the sensation. The third is the cognitive-evaluative dimension, which is about how we think of pain and the cognitive strategies we use to distract ourselves from it. This dimension includes how different cultures perceive pain, and how we can use experiences like hypnosis, music or even a pleasant taste like a sweet to control our attention and focus it on something other than the pain itself.

Illusions and synaesthesia

There's another type of pain, though, which is even less easy to control. That's the **phantom pain** people sometimes experience in limbs or other parts of the body that have been amputated. In a series of case studies, Melzack (1992) showed that these are extremely common and might even be seen as the normal outcome of amputations, whether they are deliberate or accidental. It's a particularly unpleasant example of how memory can affect the brain's current activity. Phantom pain used to be dismissed as imaginary because it obviously doesn't originate in the body. But it does influence the brain, and scans of people experiencing phantom pain show that the same brain areas are activated. So there's nothing imaginary about the pain that people are feeling. Phantom pain, it seems, originates in the somatosensory area, which contains the brain's internal body image.

We all have a body image, which seems to be prewired into the nervous system rather than learned through experience. People born without limbs often still experience whole phantom limbs, though not phantom pain. When a limb has been lost through amputation, the body image remains complete even though the body itself isn't.

Recent amputees often forget that they have lost a limb, and are vulnerable to accidents as a result. Even more disturbingly, Katz and Melzack (1990) showed how phantom limbs can carry the memory of the pain felt before the amputation. In one case, for example, a man on his way to hospital to have a painful splinter removed from underneath a fingernail was involved in an accident that crushed his arm. He still felt the pain of the splinter after the arm was amputated.

Some phantom pain responds to standard medical treatments, like analgesic drugs, or to other forms of therapy. In other cases, though, treatments may be ineffective, so the amputee has to find other ways of coping with it. Amputations are often carried out after a period of intense pain, such as from a limb being crushed, and Katz and Melzack found that anaesthetizing these painful injuries for a period of time before carrying out the amputation seems to produce much less phantom pain. They suggested that this happens because the brain has had time to recover from the pain and adjust its body image accordingly, so they recommended that this should become standard medical practice.

There are other forms of sensory imagery, and even illusions. Sometimes people experience **phantosmia**, illusory smelling, where they smell something that isn't there. It's often experienced as smelling something unpleasant – burnt or caustic, for instance – and might be in one or both nostrils. As you might imagine, experiencing phantosmia of this kind can also distort the sense of taste, and it may be caused by a number of factors, including nasal polyps or dental problems, influencing the receptor cells, or even by disturbances in the olfactory cortex.

Hallucinations can occur in any of the senses. As well as the experience of a taste that isn't there in terms of external stimulus, we may also experience a deceptive sensation of proprioception, like a feeling of lack of balance when sitting down, a sense of being touched when nothing is near us, or something similar in any other sensory mode. Although extreme hallucinations can be a sign of mental illness, mild ones, which are generally classed as disturbances rather than hallucinations, are not uncommon. Most people experience them occasionally. It is thought that they may reflect little more than random neural activity or even extremely vivid mental imagery brought on by memories. We'll be looking at imagery and memory in Chapter 7.

Sometimes, too, people experience a crossover between one sense and another, so they experience colours as sounds, or numbers as colours, or sounds as tactile sensations. This is known as **synaesthesia,** and reports of it date back a long time. In 1925

Wheeler and Cutsforth reported a case of a man who had been blind since the age of 11. Even though he couldn't see, he experienced touch and sound in terms of colours. He had done this as a child, so colours remained part of his normal experience in this way even though he could no longer see them.

Vernon (1962) suggested that synaesthesia may be the normal state for infants, and that they only gradually learn to differentiate between the different sensory modes as they gain more experience of the physical world. Other researchers have suggested that synaesthesia may result from neural confusions in the thalamus, perhaps occurring as a result of the action of neurotransmitters in neural pathways throughout that region of the brain. This idea is partially supported by the way that the drug LSD appears to create experiences of synaesthesia in those who take it. Essentially, though, we don't really know how synaesthesia develops in people who acquire it in other ways.

We do know, though, that there seem to be two types of synaesthesia. Projective synaesthesia is when the person feels that they are directly seeing colours, shapes or images in response to a particular stimulus; while associative synaesthesia is when the stimuli feel connected with other types of sensation, bringing up sensory imagery or memories. Synaesthesia can involve any of the sensory pathways in any number or combination and is generally unique to the individual, so it's hard to make generalizations about it.

Key idea

The most common form of synaesthesia is when colours are associated with visual stimuli, such as numbers, letters or musical notes. Several famous people, including Richard Feynman and Franz Liszt, have described how they experienced sensory input in this way. But there are at least 60 different types of synaesthesia: some people find, for example, that their sense of touch brings images of taste, so that the touch of an iron railing might taste salty and the touch of a piece of velvet tastes like chocolate. Other people connect touch with emotion, so oranges feel startling and silk feels calming. Some people see time as a collection of different shades and colours, with minutes, hours and weeks all having different hues; it seems that the brain can combine just about any sensation with any other. Why it does it, though, is something we have yet to discover.

Synaesthesia isn't considered to be a clinical problem: most people who experience it are quite content, and often don't realize that their sensations are unusual until they find that other people don't share them. Many people with synaesthesia see it as enriching their experience rather than limiting it, and there has been some suggestion that they may be more likely to engage in creative activities – perhaps because they are more aware of interactions between the sensory modalities.

Focus points

✳ Smell is an ancient sense, important for survival. Smell detectors are directly connected with the amygdala, which processes emotions, as well as to the cerebral cortex.

✳ Our gustatory sense makes us sensitive to many different tastes and textures. It connects with sub-cortical areas of the brain and reward pathways as well as a special area on the cerebral cortex.

✳ The sense of touch combines mechanoreception, which detects pressure of different kinds, thermosensation, which detects temperature, and nociception, which detects pain.

✳ Pain involves several areas of the brain, not just one. Feelings of social pain like embarrassment or exclusion use the same brain areas as physical pain.

✳ There are many kinds of sensory illusions, including phantom pain. Some people experience synaesthesia, where sensory images cross over or become mixed together.

Next step

We can see that our other senses all combine with vision and hearing to give us a particularly rich experience of our environment. But we don't just experience our environments passively. We are active in our worlds, and in the next chapter we'll look at how we are able to act on our environment through movement.

6

Action and skills

In this chapter you will learn:

► *how the brain controls our ability to move and balance*

► *which areas of the brain direct deliberate and automatic movements*

► *what learning athletic and other movement skills involves*

► *that performing music requires highly skilled actions involving many areas of the brain.*

We've seen how the brain receives information from the outside world and from within the body. The fundamental purpose of this is that it's evolved to help us to survive. But receiving all this information about the outside world is only useful if we are able to act on it. So we have several areas of the brain which help us do that.

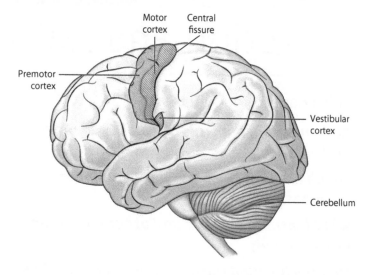

Figure 6.1 The motor areas of the brain

Moving and balancing

In order to move, we need to know where we are starting from. The previous chapters have been mainly concerned with our external senses, but we have internal ones, too, which let us know about the position of our limbs and how we are moving. **Proprioception** is one of these senses. It is the information we receive from receptors in our muscles and joints and also a special part of the inner ear known as the vestibular system. The proprioceptors themselves are small muscle spindles located in the muscles, tendons and joints. They detect position and whether muscles are flexed or contracted, and they convert this information into electrical signals which go to the brain. In this way, the brain is able to maintain a representation of the position of the body and where it is located in the immediate environment.

Proprioception is an important sense because we need to know where we are in relation to other objects in the world. It's all very well being able to see things, but unless we know where we are – how close we are to other things and what we need to do to get closer or farther away – we would be unable to act effectively on the objects in our environment. So proprioception is also closely linked with **kinaesthesia**: our sense of movement.

Kinaesthesia is an essential part of skilled action, and we deliberately train it when we learn any type of physical skill. Everyday actions like writing, drawing, getting on a bus or even making a cup of tea depend on us being able to move our muscles in the ways we want to. That in turn means that we need to be able to detect those movements with our kinaesthetic sense, so that we can correct them if we have to. Kinaesthesia is a key component of physical co-ordination and also of muscle memory.

The receptors involved in both proprioception and kinaesthesia are located in the muscles, tendons and joints. They transduce sensory information into electrical impulses, sometimes by receptors that respond if they are squeezed or stretched by muscle movement, and sometimes by receptors which can detect passive position, like the angles involved in the position of joints. These impulses then travel to the central nervous system, taking one of two distinct pathways. One pathway is active in conscious movement. Neurones following this pathway travel to the base of the medulla, where their messages are picked up by other neurones and passed to the thalamus. Other relevant sensory information, such as vision, links up with it at this point, and then the information goes to the area at the very bottom of the somatosensory area of the cerebral cortex – an area known as the vestibular cortex.

The other pathway is activated by unconscious, or automatic, movement, and it ends up in the cerebellum rather than the cerebrum. Information from the proprioceptors travels first to the spinal cord and then through the medulla and the pons, before ending up in the cerebellum. As we saw in Chapter 1, the cerebellum is the part of the brain that co-ordinates skilled movement, so it needs to receive information about proprioception and kinaesthesia.

BALANCE
Both proprioception and kinaesthesia also involve the sense of balance, which is sometimes called **equilibrioception**. We achieve and maintain our balance as our muscles react to information about our bodily movement and orientation, making slight adjustments to it as necessary. The information comes from three sources. Some of it comes from our vision: have you ever been in one of those 'crazy

house' illusions which make you feel as though you are on a slope when you aren't? By giving your eyes the wrong information, it can really affect your sense of balance. Other information about balance comes from proprioception: if you stand on one leg and reach upwards, you will feel the muscles in your leg and foot making small movements to keep you upright. The brain is using that proprioceptive information to keep you balanced and steady.

We also have a special system for detecting which way up we are, and whether we are being moved. This is the third source of information, and it is detected by a part of the inner ear known as the **vestibular system**. The inner ear is filled with fluid, and part of it contains the organ of Corti, which as we saw in Chapter 4 is how we detect sound. But it also contains three structures known as the semicircular canals. These are fluid-filled loops at roughly right angles to one another. The fluid responds to gravity, so there is different pressure along different parts of the canals. It makes a kind of 'push–pull' system: when one part of a canal has more pressure, its matching canal in the other ear will have less. By comparing the inputs from both ears as the fluid swirls around in response to movement, the brain can identify how the head is moving or rotating.

It is the movement of the fluid in the semicircular canals that produces the sense of dizziness you feel if you have been spinning and suddenly stop. Some performers, like ballet dancers, control this by moving their head very rapidly as they turn, and then stopping it for longer in the forward position, to minimize the disturbance of that fluid. Other performers, like ice skaters, couldn't do that, so they have to learn to skate on anyway, ignoring the dizziness and trusting their muscles to perform their well-practised actions.

The semicircular canals also detect linear movement, by means of sensitive membranes which contain tiny particles known as **otoliths**. If your head is stationary, the otoliths press downwards on the membrane, but if you are moving forward or in another linear direction, the otoliths press backwards or sideways. The pressure is detected by hair cells, stimulating them to produce an electric impulse. Information from these two sources goes to the medulla and then on to the cerebellum, where it links with visual input to produce our sense of balance. People who feel motion sickness often find it is worse if they try to read or do something involving static vision. It might not seem to make much difference, but often looking out of the window reduces the nausea a little, because the brain is more able to make sense of what it is feeling.

Anti-motion sickness drugs usually act directly on the cerebellum to suppress this activity.

Taking action

Receiving information is only part of the story: we also need to be able to act on the information we receive, and that means being able to move – to perform actions. Imagine that you've been sitting down for some time and you decide you want to go to the kitchen to get something to drink. It all starts with the frontal lobes – the forward part of the cerebrum immediately in front of the central fissure and above the lateral fissure. The frontal lobes are all about control of what we do. At the very front they are concerned with aspects of thinking, like planning, reasoning and making decisions, while the farther back we go, towards the central fissure, the more they become concerned with actions and movement.

The very front part (the anterior) of the frontal lobe is, as we said, generally concerned with decision-making, thinking and other cognitive activities, so that's the part of the brain which becomes active when we decide to do something. At the very back of the frontal lobe is a strip of cortex that runs alongside the central fissure, opposite the somatosensory area (see Chapter 5). This is the **motor cortex**, which is the part that sends direct messages to the different parts of the body so that they can move. But there are several steps between deciding to take action and actually carrying out that movement, and those steps are reflected in the areas of the frontal lobes in between.

Key idea

One of the more exciting developments of this century has been the way that neurologists and cyberscientists have been able to work together to develop effective working prosthetics for people who have experienced significant injuries that have left them without the use of their limbs. For example, sometimes people who have become tetraplegic (both arms and legs paralysed) as a result of a spinal injury still have relevant neural activity in the motor cortex. Some people have been able to train their brains to use impulses from that part of their brain to move a curser on a computer screen. More recently, some people have used these cortical signals to operate specially designed robot hands or limbs. It takes considerable practice, but in many cases it gives these people a new life.

Just behind the anterior frontal lobe is another area, known as the **prefrontal cortex**. This area is also involved in what we think of as higher mental functions like working memory, thinking and decision-making, and it is here that planning takes place. This part of the brain isn't concerned with specific movements but with the goals of the action – for example, the aim of doing something about feeling thirsty. In one study, for example, people were asked to move a finger, either in response to being touched or as a result of their own decision about which finger to choose. Both actions activated the parts of the frontal cortex concerned with specific actions, but the second task also produced activity in the prefrontal region, showing how it is involved in making choices and having intentions.

In another study, teenagers were asked to manipulate joysticks while their brain activity was scanned. They were given two conditions: one in which they moved the joystick in response to an auditory signal and another in which they made their own decisions about when and how they would move it. While both conditions involved activity in the premotor area and the motor area, only the second condition showed activity in the prefrontal cortex. The brain scans reflected their decisions about the nature and timing of the movements – that is, what they were going to do and when they were going to do it– reinforcing the idea that this part of the brain is directly concerned with the planning and decision-making involved in deliberate movement.

If their prefrontal cortex is damaged, a person often shows disorganized or inappropriate behaviour. They may, for example, carry out familiar activities unnecessarily like switching lights on or off, or opening and shutting cupboards at random, or they may repeat actions over and over again. They carry out automatic actions or action sequences, but what they seem unable to do is co-ordinate their thoughts and intentions with their activity. They may also become very suggestible, responding readily to proposals for action from other people and becoming easily hypnotizable. Essentially, if we make our own decisions about our actions, we do it using the prefrontal cortex. But if we are simply responding to external demands from other people or acting automatically because of the situation, we don't.

Behind the prefrontal cortex and in front of the motor cortex is an area known as the **premotor cortex**. This is the part of the brain that prepares the motor cortex for action and adjusts which neurones it will be using. The part of the premotor cortex towards the centre of the brain is sometimes called the **supplementary motor area,** or SMA, and it receives proprioceptive information about how parts of the body are positioned. The SMA is involved with well-learned actions, like playing a familiar tune on an instrument or typing on a keyboard – in other words, actions that don't require the brain to be constantly checking the external environment for new information. This area of the brain seems to prepare well-learned motor sequences of movements that have to be executed with precise timing. It does this by combining feedback from the proprioceptive receptors in the muscles with feedback from other sensory areas, like hearing or vision. So it is very important for musical performance (we'll come back to that later), and it's active in other well-learned actions as well.

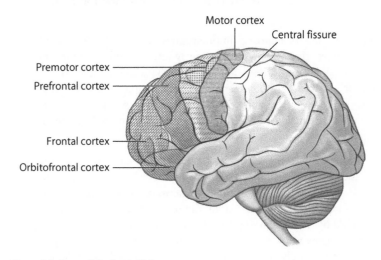

Figure 6.2 Areas of the frontal lobe

The supplementary motor area is the part of the premotor cortex at the top and centre of the brain. Farther down the side of the brain, the premotor cortex has a different name – the **lateral premotor cortex** – and a different function. It becomes active when we are acting on objects in the environment, in other words in situations where we need to be scanning the environment for external information, like reacting to a traffic light or opening the microwave door when it pings. The lateral premotor cortex receives some information from the other senses, but it responds most strongly to the visual information coming from the parietal lobes – in other words, visual information which has been processed for its meaning and implications. It is the lateral premotor cortex which is involved when you fire a weapon in response to the sight of an 'enemy' in a video game, or hit a button to indicate that you know a quiz answer, or slam on the brakes if you see a deer about to walk into the road.

One of the more fascinating findings made since the 1990s concerns **mirror neurones**. These were originally thought to be specialized nerve cells that respond both when we perform actions ourselves and also when we see the same movements in other people. Since those early findings, though, we have found that mirroring isn't about specialized neurones. Instead, the same neurones we use to represent our own actions are triggered when we see other people doing the same. Watching someone else move their arms in a particular way generates the same neural activity in the premotor cortex which would be involved if we ourselves were moving our arms like that – only sometimes not quite as strongly.

There are mirror neurones in other parts of the brain, too, and the whole mirroring system is particularly interesting because it reflects the social nature of the human being. What other people do is important to us, and understanding their actions gives us a way of responding appropriately to what they do. We will come back to this idea again as we look at the social aspects of the brain in later chapters.

To come back to the question of action: the premotor cortex sends messages directly to the area on the top of the brain, at the very back of the frontal lobes – the motor cortex. When neurones in the motor cortex are stimulated, the relevant parts of the body respond by movement. This is because some of the neurones in the motor cortex send messages directly to the thalamus, which passes them immediately on to the muscles, stimulating the muscle fibres to contract. Like the somatosensory cortex, the motor cortex is organized according to the different areas of the body. The feet are at the top, followed by the torso, then the hands, which have a large area of cortex because we use our hands in so many ways, then the head and face, which also has a large area for the same reason, and finally the tongue (see Figure 6.3). There's a crossover here: as with the somatosensory cortex, the motor cortex on the left frontal lobe controls the right side of the body, while the motor cortex on the right frontal lobe controls the left side of the body.

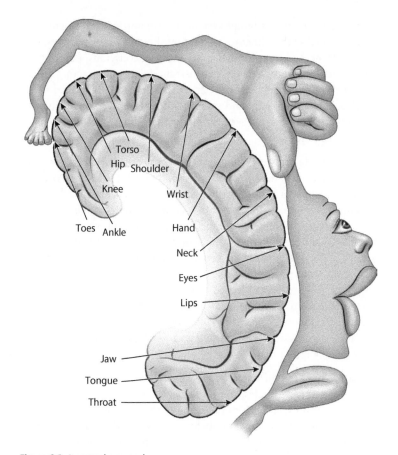

Figure 6.3 A motor homunculus

That's all well and good, but how does the brain decide which neurones should fire? Researchers have found that the pattern of neural activity in the motor cortex is all about the direction of movement. Each neurone in this area has a 'preferred' direction, and is most active when the movement is in that particular direction. But they also fire, less strongly, if the action involves movement in a similar direction. The movement itself – that is, the set of electrical impulses which will finally trigger muscle contraction– is the result of a combination of the number of neurones which are firing, and how strongly each one fires.

MOVEMENT SYSTEMS

Some of what we do consists of deliberate, planned actions. But many of our actions don't involve a high level of awareness: they are carried out automatically without our conscious involvement.

So the brain has two different systems for organizing movement – one concerned with deliberate movement, known as the **pyramidal motor system,** and another that is all about unconscious, automatic types of actions, known as the **extrapyramidal motor system.** Both of these systems involve the areas we have already looked at – that is, the prefrontal and premotor cortex, and particularly the supplementary motor area. But after that, their messages go in different directions.

In the pyramidal motor system – the one concerned with choice and conscious actions – the messages go directly from the motor cortex to the muscle fibres via the thalamus. This is how we respond with movement if the relevant part of that area is stimulated, and it's also how researchers were able to build up the homunculus map of the motor area shown in Figure 6.3. Some people who have suffered injuries or strokes have learned to use this area to direct computer cursors, or in some cases even robotic arms or hands. It's a relatively new area, because the electronics are so complex and the people concerned need to put a lot of effort into learning how to do it, but essentially it involves the person imagining the specific movement. The nerve signals that are activated as a result are then used to direct the computer or robot.

Messages that take the extrapyramidal route through the brain also end up at the muscles, but less directly. Instead, they follow one of two routes, either looping through the basal ganglia or looping through the cerebellum before they go to the thalamus and then to the muscles. The basal ganglia are large, rounded masses surrounding the thalamus, which form part of the limbic system of the brain. They are particularly active in timing our movements, which is important for co-ordination. Studies which ask people to do timed finger tapping, for example, show increased activity in this area of the brain. Damage to this area of the brain can result in tremors and a lack of muscular co-ordination, as seen in people suffering from Parkinson's disease.

The basal ganglia also receive information from many other areas of the cerebrum. Movement information which loops through the basal ganglia connects with information coming from the sensory areas of the cortex – information about vision, hearing, cognition and proprioception, for example. From there, the information (or most of it, anyway) goes on to the thalamus, which directs the messages that go to the muscles. That's the route taken by our unconscious reflexes, like ducking if we see something flying through the air towards us.

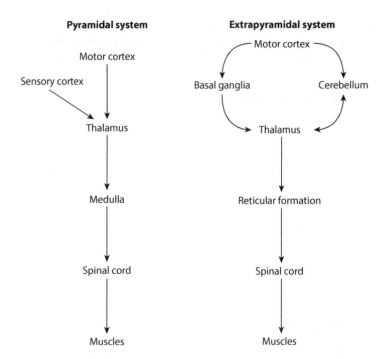

Figure 6.4 The brain's two motor systems

The route that loops through the cerebellum is also concerned with actions that we do without thinking. Well-learned actions and fluid movements are controlled by the cerebellum, which stores automatized action sequences and motor programmes to produce smooth, skilled actions. A skilled keyboard operator, for example, doesn't think about where they are placing their fingers; an experienced gamer moves a joystick and other controls without thinking, and a well-practised driver doesn't think about the actions involved in changing gear. Even if a driver makes a conscious decision to change gear, the muscle actions themselves are performed in a smooth, well-practised sequence. A novice driver, on the other hand, doesn't have those stored action sequences in the cerebellum, and so has to think about everything they are doing, making driving feel like a very complicated process.

Damage to the cerebellum can mean that the person moves only tentatively: they can move but they lack balance, and they also lack the confidence that comes from being able to move smoothly and without thinking. Unlike the motor cortex, the cerebellum is ipsilateral – that is, the left side of the cerebellum controls the left side of the body and the right side of the cerebellum controls the right side of the body. If it is damaged on one side, the person is usually still able to move, but that side of the body is likely to be badly co-ordinated, producing staggering, slurred speech or similar problems.

Case study: Alien hand syndrome

One of the stranger symptoms that can result from brain damage is known as alien hand syndrome. In this, the person concerned has no sensory feedback from one of their hands – often the left– and it seems to take on a life of its own. Records of people with this syndrome include a man whose left hand would pull his trousers back down when he pulled them up with his right hand, a card player whose left hand refused to let go of a card she was playing, and even a woman whose left hand tried to strangle her one afternoon. Autopsy and fMRI research indicates that this results from damage to the right primary motor cortex and also other areas of the frontal lobes that deal with intentions and the planning of actions.

Learning skills

How do we develop the smooth action sequences that we take for granted in our everyday lives? Effectively, it's by training groups of nerve cells to become used to one another. Back in the 1950s, long before brain scanning had become a reality, D. O. Hebb proposed that learning happens because cells form what he referred to as cell assemblies. The more often one neurone stimulated another, the stronger its connection would become: Hebb believed that the synaptic knobs making the connection between the cells would grow larger, so that their effect was stronger. As we saw in Chapter 2, modern research shows that he was very nearly right: the synaptic knob itself doesn't grow but the number of vesicles and receptors associated with that synapse increases the more the connection is made. It's known as **synaptic plasticity**, and it means that the next nerve cell gets a stronger message when the first one releases its neurotransmitter, making it more likely to fire.

This is why practice is so important. Each time we practise a sequence of actions, we are making our brain cells follow one particular sequence of connections rather than any other. The more often we do this the easier that particular sequence becomes, because those neurones will fire more easily. There are other ways that practice makes a difference to neurones, too. Bengtsson et al. (2005) found that practice in piano training increases the amount of myelination in brain fibres, particularly in the frontal lobes and the corpus callosum. As we saw in Chapter 2, myelination makes the impulse travel faster along the neurone, so increasing the amount of myelination helps action sequences to happen more quickly.

As any athlete will tell you, learning a new skill isn't just a straightforward process of getting better all the time. Instead, learning proceeds in a series of learning curves, broken up by plateaus. Figure 6.5 shows what this looks like. At first, each practice session or set of practice sessions produces a noticeable improvement. After a while, though, it levels out and there doesn't seem to be much improvement at all. Although people can sometimes find this dispiriting, it's a time of consolidation, where the brain is integrating those movement programmes into its repertoire of actions. Eventually, the improvement starts again and this time the skill is more solidly based. Improvement then continues until the next plateau is reached. This means that if you're trying to learn something and were getting better but don't seem to be improving any more, it's important to keep practising. It's worth it!

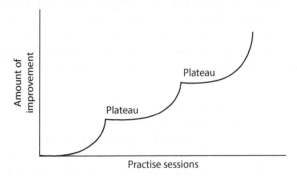

Figure 6.5 Plateau learning

Practice can be mental as well as physical. Studies have shown that mentally imagining actions can produce very slight changes in the strength of the muscles concerned – not nearly as much as physically doing the actions would, of course, but enough to register as a difference in physical tests. As a result, training in mental imagery is an important part of a professional athlete's training. Mentally rehearsing the physical actions involved in successful performance helps the neurones to consolidate their connections, and strengthens the relevant neural pathways.

NEURAL PLASTICITY
There is some evidence, too, that the body is able to regrow neurones, which is good news for people who have suffered injuries

damaging neural connections to the limbs. Neural regrowth is more likely in the peripheral nervous system than in the complexities of the central nervous system: as a general rule, the brain tends to respond to injuries by rerouting the neural connections and sometimes making use of nerve cells which were involved in other, less important functions. But in the rest of the body it seems to be possible, as long as the nerve cell body is intact and in contact with some Schwann cells, which provide the relevant chemicals. What happens is that the nerve cell produces a growth cone at its severed end, which gradually rebuilds the axon down towards the limb. This regrowth can be quite rapid – as much as 5 millimetres a day in the larger nerves – but it will be impeded by scar tissue and may need to be helped by chemicals that stimulate nerve growth.

It also seems to be directly affected by effort: the more the person tries to move the affected part, the more the nerve cells seem to respond. Although this isn't very well understood, we know that the brain is sending movement signals to the muscles, and it may well be that stimulation of that kind at the beginning of the nerve cell helps to stimulate the neural growth at the other end. Some people have shown quite amazing recoveries from serious injury as a result of making determined and sustained efforts, guided by therapy, over time; while others who have suffered similar injuries remain damaged or even paralysed for life.

Other aspects of movement can also influence neural plasticity. There is increasing evidence that exercise at younger ages can have protective effects on brain functioning as people get older. These are mostly retrospective studies, of course, and we won't be certain of this effect until we have been able to follow people through their lifespan. But a number of studies have shown how exercise in middle life, as mature adults, can also help general mental functioning, and also seems to reduce people's likelihood of developing degenerative brain diseases. Raichlet et al. (2016) scanned the brains of young endurance athletes. When they compared them with healthy individuals of the same age who didn't exercise, they found significantly more connectivity across many different areas of the brain – not just those involved in visual and motor functioning but also areas concerned with working memory and cognition.

Our neural mechanisms of movement, then, are not simply wired in. They are constantly responding to practice, learning and the challenges we give them. Through practice, we can develop physical skills and abilities that go far beyond what we thought we would be capable of. One of the connections between brain and movement that we are coming to understand fairly well is the relationship between brain activity and musical performance.

Performing music

In musical performance, the movement itself is produced by muscle contractions, but exactly which muscles contract, and when, is a much more complex matter. The advent of brain scanning, though, means that we can ask musicians to play their instruments while their brain is being scanned, and so we can get an idea of which areas of the brain are involved and in what sort of sequence. In an extensive review of studies of brain functioning and music, Zatorre, Chen and Penhune (2007) described how musical performance involves three basic aspects of physical control: timing, sequencing and spatial organization. Each of these involves different brain areas.

Timing, of course, is essential in musical performance, and it seems to involve three main areas: the cerebellum, the basal ganglia and the supplementary motor area (SMA) of the premotor cortex. Research

Figure 6.6 Musical performance

using PET scans and other sources suggests that the cerebellum is involved in precise short-interval control, measured in milliseconds, while the basal ganglia and supplementary motor area are more important for longer time intervals, measured in seconds. We saw in Chapter 4, though, how our sense of hearing has neural pathways directly concerned with timing of sounds and rhythmical activity, so those are not the only brain areas concerned with timing (although they do seem to be the main ones). Performing and responding to complex rhythms involves other areas of the cerebral cortex as well.

We have also seen how the premotor cortex, particularly the area known as the SMA, has a role in preparing the motor cortex to carry out timed sequences of actions. Timed sequences, of course, are an essential part of musical performance, so this part of the premotor cortex is particularly active in performing. But so is the lateral premotor cortex, which you may remember becomes active when responding to signals from the environment. Performing musicians are alert to the feedback they are receiving from their own actions, and they are also alert to information about the actions of others, particularly when they are performing together. So both areas of the premotor cortex are involved in sequencing.

The cerebellum is also involved in sequencing; its ability to respond in milliseconds means that it can control movement trajectories in a very precise way. We saw earlier how the cerebellum receives and integrates inputs from many different senses. It uses this sensory information to develop predictive models of the likely results of actions, and uses these models in both feedback and feedforward – that is, in the rapid correcting or adjustment of actions in response to errors and in dealing with what might become errors if they are not adjusted quickly. By combining proprioceptive, auditory, tactile and visual information, the cerebellum produces the smooth action sequencing which is so important for skilled musical performance.

Many studies have shown how the cerebellum is involved in learning sequences of actions, and linking those action sequences to form larger units. It is noticeable that when musicians are practising a piece that they are learning, they don't correct individual notes if they make mistakes, as novices sometimes do. Instead, they go back and repeat whole sequences. This emphasizes how important the sequencing of actions is. It encourages the cerebellum to integrate whole chunks of a tune or piece of music to create larger units of skilled performance. A similar (though less sophisticated) type of learning happens in typing, where the sequences relating to whole words are stored and performed, rather than each individual letter being consciously touched one after the other.

Brain-scanning studies have also shown how the basal ganglia are involved in this aspect of music performance. Connections between the frontal cortex and the basal ganglia become particularly active while movement sequences are being learned, but they are also active when well-learned sequences are being performed. In one study (Bengtsson et al., 2004), people's brain activity was scanned while they were asked to tap out complex rhythms using the right index finger. Sometimes there was only one key involved but they were required to produce a complex rhythm, emphasizing the need for control of timing. In other trials, there was a simple rhythm but the person had to tap several keys, emphasizing the need for sequencing. The researchers found that different brain areas were involved in the two types of task. The basal ganglia, the cerebellum and the side of the cerebrum involving both frontal and parietal lobes were more active during the sequencing task. The premotor area and parts of the temporal lobe were more active in the timing task. So musical performance, which demands both timing and sequencing, combines activity in all of these brain areas.

There has been relatively little research looking explicitly at how spatial organization is involved in music performance. Some brain-scanning studies, though, have found that the part of the premotor cortex nearest the motor cortex is particularly involved with spatial learning. A study of expert cellists showed that they had an unusually high level of accuracy in placing their fingers, by comparison with most other people. Normally, we see a trade-off between accuracy and speed – the faster someone performs an action, the less accurate they become. But expert string players have to be accurate in their placing, because the slightest variation will affect the note they produce, and they don't show the speed-accuracy trade-off at all. They place even very rapid movements perfectly.

REPRESENTING MUSIC
Timing, sequencing and spatial organization may be the elements of musical performance, but trained musicians also process information in other ways. Studies have shown, for example, that they are much more likely to use kinaesthetic representations when they are remembering musical pieces. While an untrained person may simply remember the sound, trained musicians are also likely to be remembering the feel of the muscle movements involved in producing it. These kinaesthetic representations may act as a kind of mental practice, which helps the musician concerned to consolidate their skill learning in much the same way as mental imagery can help in athletic performance or physical skills.

Skilled musical performance often involves a high level of creativity as well. Liu et al. (2012) performed brain scans on freestyle rap artists as they engaged in two tasks. The first was to produce a spontaneous improvised freestyle rap, and the second to perform a well-rehearsed set of rap lyrics. The performers showed similar activity in the motor cortex and the premotor cortex for both tasks. However, the improvised performance showed much more activity around the supplementary motor area (the medial prefrontal cortex) and in the language areas, reflecting the need for the performers to be able to select words rapidly, both for rhythm and meaning. In fact, there was more activity right across the left hemisphere in general for the improvised performance. When they performed the rehearsed lyrics, by contrast, there was more activity in the lateral prefrontal cortex but not in other areas of the brain. As we've seen, the lateral prefrontal cortex is concerned with movement produced in response to external demands. Both of the tasks in this study involved the use of language, but the improvised task showed more activity in the language areas of the left hemisphere than the rehearsed task, which had more to do with remembering.

Focus points

* The bodily senses of proprioception, kinaesthesia and equilibrioception (balance) all help us to know what our bodies are doing and to move effectively.

* The premotor cortex plans movement and the motor cortex directs deliberate movement through the pyramidal motor system. The extra-pyramidal system involves rapid survival reflexes without cortical control.

* Skill learning involves the transfer of movement control from the motor cortex through the cerebellum, as a result of practice.

* We can sometimes regrow neurones or train areas of the motor cortex to use new pathways when recovering from stroke or injury, but it requires sustained effort.

* Musical performance requires highly skilled actions performed accurately and in sequence, and it involves the basal ganglia, the premotor cortex and the cerebellum as well as the motor cortex.

Next step

We will be looking at language processing again in Chapter 10, but in the next chapter we'll go on to look at the brain areas involved in memory.

7

Remembering

In this chapter you will learn:

- ▶ *the difference between working memory and active thinking*
- ▶ *how we remember events and contexts*
- ▶ *how the brain stores new information and maps*
- ▶ *what happens in forgetting and amnesia.*

One of the first things you'd be likely to say if someone asked you what your brain is for would be: 'Remembering things'. And that's quite true. It could even be argued that memory is what the brain is all about. We know from studies of neural plasticity that all areas of the brain are able to adjust themselves and learn from experience, and that process in itself is a kind of memory. But there are other kinds, too, and what we usually think of as memory actually involves several different processes, which engage areas all over the brain. For example, we can represent information in many different ways. We store some memories using images – sensations like tastes, smells and visual images. We can also store memories verbally, using language, or symbolically, using other types of concept.

We have different kinds of memory, too. We have our long-term autobiographical memories of things we have done and what has happened to us in the past, which psychologists call episodic memory. We have memories of other things, which haven't affected us personally but are still information that we know – semantic memory. We have memory that is less about stored information and more about how to go about doing things, which is known as procedural memory. And we have prospective memory, which is all about planning and remembering to do things in the future.

Most of our memories remain unconscious until we need them. They come back to us if they are cued in some way – in other words, if something reminds us of them. We might recognize something or it may seem familiar, even if we couldn't remember it without those hints, and that's a type of memory, too. And even if we are not aware of them, we draw on our memories to solve problems and make decisions. Psychologists know a lot about the different types of memory – they've been studying the subject since the nineteenth century (my book *Understand Psychology* tells you more about it), but here we are more interested in the brain's activity and how memory works in the brain. We don't know all the answers, but we do know quite a lot.

Working memory and active thinking

When we are thinking and working out problems, we use a kind of memory called **working memory**, which is how we keep things in mind while we think about them. Working memory is very short term, and often involves new information that hasn't been stored long term. That's limited, though: studies have shown that we can't hold much new information at a time, and that new information is often forgotten just as soon as the brain has finished what it was doing with it. That's why the memory store on your mobile phone is so important: you might remember someone's phone number while you put it in for the first time, but then you forget it straight away. If the phone didn't do the 'remembering' for you, you'd have to look up the number separately each time you wanted to use it, and you might not be in a position to do that.

The very front part of the brain – what's known as the anterior frontal cortex of the frontal lobes – is the part of the brain most actively involved with conscious thinking (see Figure 7.1). It doesn't deal with remembering as such, although it draws on our memories while it is dealing with other cognitive matters. But immediately behind that area are two other areas: the dorsolateral prefrontal cortex at the top and the ventrolateral prefrontal cortex below it, and these are actively concerned with working memory. Incidentally, the 'lateral' part of their names refers to the way that they are both at the side of the cerebral hemisphere rather than in the middle, and the 'dorso' and 'ventral' parts mean 'towards the top' or 'towards the bottom'. So the dorsolateral prefrontal cortex is the part of the cortex near the top but on the side, and just before the frontal cortex itself. The names sound complicated but all they do is indicate where that part of the brain is.

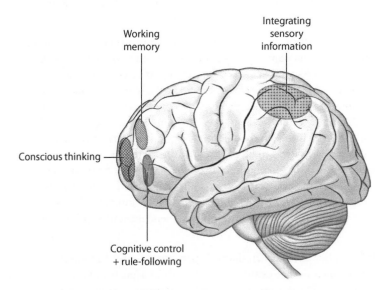

Figure 7.1 Memory in the frontal lobes

The **dorsolateral prefrontal cortex** (PFC) is involved in working memory. It becomes more active when we are asked to recall the context of a memory – like which holiday it was when we first saw a famous landmark – rather than just, say, recognizing the landmark itself. It becomes active if we are asked to group memories into meaningful sets ('How many different types of transport can you remember?'), but not in random free-recall tasks ('Tell me what you are remembering right now.'). It becomes more active if it is presented with strings of digits (numbers or letters) with some structure to them, like numbers that follow a particular sequence, but not when we see sets of digits with no connection. So essentially, this area of the brain becomes active when we are organizing information or exploring the relationships between things.

Laboratory studies involving scans of the dorsolateral PFC have shown that it is also involved with uncertainty. Most of us are familiar with the **tip-of-the-tongue phenomenon** (usually shortened to TOT), in which we know that we know something and we can nearly remember it but not quite. Brain scans of people experiencing TOT show a lot of activity in the dorsolateral frontal cortex of the right hemisphere. That activity isn't there if they are sure they don't know the answer or when they are confident that they can remember it. What all this adds up to is that the dorsolateral frontal cortex becomes active when we are actively thinking about, evaluating and manipulating information.

The other part of the prefrontal cortex, the **ventrolateral prefrontal cortex**, is (as its name suggests) underneath the dorsolateral PFC. It seems to be mainly concerned with cognitive control and rule following. Many of our activities, particularly in the modern world, demand that we are selective in what we pay attention to. Driving, for example, requires constant visual control and attention on the part of the driver – you would be a very dangerous driver if you allowed your vision to be continually distracted by other things around you – and many other activities have 'task rules' of this kind. That's what researchers mean by cognitive control, and it's an important part of working memory. If we don't stay focused on a problem, we forget its important elements and won't reach a sensible solution.

PET studies have shown that this part of the brain becomes active when we are dealing with the spatial aspects of working memory – that is, tasks involving locations or putting things in particular places in order to solve problems. Damage to the ventrolateral PFC affects how well we are able to co-ordinate muscle actions (like the eye muscles we use to focus clearly) with the demands of working memory and cognitive monitoring. So this area is all about what we are actually doing with our memories, in the sense of integrating them with our actions and thoughts.

Information that is going to be stored for a longer time has to be consolidated, so that it becomes integrated with the rest of what we know. The ventrolateral PFC also seems to play an important part in establishing new memories. The dorsolateral prefrontal cortex does this too, but it can only do it with structured information, not random facts. The ventrolateral PFC can sometimes help us to store unstructured information, but how successful it is depends on how much cognitive processing is involved. Psychologists have known for a long time that the more the brain processes information, the more likely we are to retain it in long-term storage: we are more likely to forget random or unstructured information.

In Chapter 6 we saw how other parts of the prefrontal cortex become active in response to our intentions – when we make a decision to act or when we plan an action. The ventrolateral and dorsolateral PFC work together to co-ordinate the working memory involved in that sort of planning, linking with other parts of the prefrontal cortex in the process. Each of these areas of the frontal lobes also draws information from the parietal lobe of the brain – the area towards the back of the brain, at the top, which integrates various types of sensory information, linking vision, hearing, smell, touch, proprioception and so on.

In doing this, the parietal lobe feeds all that information into a pathway that makes multiple connections with areas in the frontal lobes, including the ones we've just been looking at. That pathway is often called the dorsal visual stream, but the name is a bit misleading because, as we've seen, it involves much more than just visual information – as do our memories. We'll come back to the idea of imagery in memory later in this chapter.

Remembering events and places

The frontal lobes, then, are to do with how we use our memories to help us think. But that leaves the question of how we store our memories in the first place, and to understand that we need to look at what is happening underneath the frontal lobes, in an area known as the **medial temporal cortex** – the bit of the temporal cortex tucked away right in the middle of the brain.

The temporal lobes, as we've seen, are at the side of the brain, and are folded round into the lateral fissure, so they continue underneath it. At the very edges of those lobes, right next to the subcortical structure known as the hippocampus (about which more later), are two areas of the temporal cortex. They are not visible from the outside, because they are hidden deep in the brain on either side of a deep groove known as the **rhinal sulcus**. The area on the same side of the groove as the hippocampus is called the entorhinal cortex, and the area on the other side is the perirhinal cortex. Next to the perirhinal cortex, and curving round so that it lies next to the hippocampus, is a third area known as the parahippocampal cortex (see Figure 7.2). These areas are so tightly connected with the hippocampus and one another that, although they are all part of the cortex, they are sometimes thought of as subcortical areas rather than directly part of the temporal lobe itself.

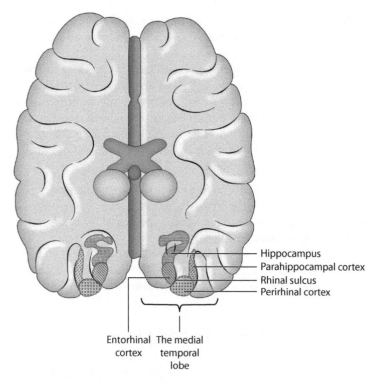

Hippocampus
Parahippocampal cortex
Rhinal sulcus
Perirhinal cortex

Entorhinal   The medial
cortex   temporal
            lobe

Figure 7.2 The medial temporal cortex

The **entorhinal cortex** seems to be the main interface between the cortex and the hippocampus. Its primary job is transferring memories between them, so it's a crucial element in how the brain is able to use its memories. It deals with episodic and semantic memories, acting as a major focus point for neural pathways involved in memory and navigation. It doesn't deal with procedural memory, though. People with damage to this area are still able to interact with their environments and do the everyday things they could do before. But their ability to store new memories becomes severely limited.

Single-cell recordings in the entorhinal cortex have identified specific cells that respond to where we are located in the environment. There three types of these cells: grid cells, path cells and speed cells. Grid cells seem to be arranged in hexagonal patterns and are thought to represent places in a cognitive map. Path cells respond to directions and orientation. And speed cells respond to the rate at which the movement is taking place. Interestingly, though, these cells are not responding to the physical action itself but to the movement we perceive ourselves to be making. In normal life the two are usually the same, but not in virtual reality. Single-cell recordings of people playing video games have found path cells that respond differently depending on whether the person sees themselves as moving in a clockwise or anticlockwise direction, regardless of where they are in their perceived location. And these cells act in the same way regardless of whether it is the player themselves who is moving or their avatar moving around in a virtual space.

On the other side of the rhinal sulcus is the **perirhinal cortex**. This part of the brain is also directly involved in recognition and feelings of familiarity. It codes whether objects are familiar or unfamiliar, sending that information to the hippocampus as well as to other areas of the brain, and it is also involved in the perception of complex objects. It helps to process representations of things and makes connections between different types of stimulus. It can do this because it has strong links with the visual perception and other sensory pathways, as well as with the memory pathways in the brain. All this means that it is important for establishing the meaningfulness of objects or items.

Many people have a strong memory for locations and places. They can remember where they are even if they have visited a place only once, many years previously. This is possible because of the third of these areas, the parahippocampal cortex, which feeds information into both the entorhinal cortex and the perirhinal cortex. It, too, plays an important role in coding memories but, whereas the perirhinal cortex is about coding representations of items, the parahippocampal cortex is concerned with the representation of contexts, including how we perceive scenes and landscapes. It communicates with the hippocampus to code these memories into long-term storage, and also becomes active when we are retrieving these types of memories.

A particular group of cells in the parahippocampal cortex is directly concerned with identifying and recognizing specific places. This is known as the parahippocampal place area, or PPA for short. Studies using fMRI have shown how this area of the brain becomes active when people are looking at images of landscapes, rooms or groups of buildings. Participants in other studies that have stimulated this area electrically have talked about seeing places or scenes, often vividly enough to be almost a hallucination. Incidentally, there's another area, very close to this one, which responds in a similar way but specifically to faces. We'll be looking at that in Chapter 9.

The contexts of our actions can be social as well as physical. There is some evidence that the parahippocampal cortex responds to the social context of our interactions – particularly the social context generated by different ways of using language. We all use many different language registers: we can speak with others in a range of ways including very formal, friendly or excited. Each of these ways of using language generates a different social context or 'feel' to the social interaction, which seems to be interpreted by this area of the brain. For example, an area in the right parahippocampal cortex becomes particularly active when the person is responding to sarcasm. In order to detect sarcasm, we need to be aware of what the other person is thinking, or at least aware that what is being said is a put-down and not meant to be taken literally. As social animals, we can see how this type of context would be as important for selecting appropriate action as the physical context.

Storing new information and maps

The three areas we have just been looking at all surround the **hippocampus** and feed information into it. The hippocampus is a seahorse-shaped structure (hence its name) sometimes described as part of the medial temporal lobe, but tucked right underneath it. It is the main centre for the consolidation and storage of memories in the brain. The hippocampus integrates the information from all three of the areas we have been looking at, and processes it for long-term storage. People with serious damage to the hippocampus become unable to store new memories, while those with lesser damage also find themselves with difficulties in that area, which may be general or only concern specific information, depending on the damage.

One thing the hippocampus does is link items of information to their context. This generates meaningful 'mental maps', which we can use to move around environments. The hippocampus contains place cells, which fire only in response to a given location and context. These place cells combine their information, creating an allocentric representation of a place or places (allocentric means independent of the person's actual location, as in a map). In a classic study Maguire et al. (2000) found that London taxi drivers had an enlarged right hippocampus by comparison with other people. A follow-up study in 2011 ruled out other possible explanations and showed that this enlargement was directly concerned with the demands of the job, which require a powerful knowledge of locations and routes.

It is spatial memory, and in particular an extensive cognitive map of London, which forms 'the Knowledge' required of all London taxi drivers, so it is the right hippocampus rather than the left one that is affected by their occupation. The 2011 study by Woollett and Maguire was a longitudinal study which showed that the longer the drivers spent in the job, the larger the right hippocampus became. They also studied a taxi driver who had an accident resulting in damage to both sides of the hippocampus. He did retain a broad knowledge of London streets, in terms of the main roads, but lost all his detailed knowledge of the side roads and less common routes.

The hippocampus on the right side of the brain is mainly concerned with spatial memory, including large-scale allocentric maps of the environment, while the left hippocampus appears to store other contextual details. In one study, people with hippocampal damage were given a task that involved learning their way through a virtual town. They were given objects at specific locations along the way, and when they finished they were asked to draw a map and identify the objects and scenes they had encountered. People with damage to the left hippocampus had trouble remembering the objects but could draw the map reasonably well, while those with damage to the right hippocampus recognized the objects but had problems recognizing scenes and drawing the map.

Looking at how these areas work together, we can see how the perirhinal cortex and parahippocampal cortex process content and context separately. The perirhinal cortex processes representations and the parahippocampal cortex processes context, while the hippocampus binds them together, placing item representations in context and consolidating them in long-term memory. There are different ways of remembering as well as different types of memory, and it has been suggested that the first two are important for familiarity and recognition, while the hippocampus is important for recollection – that is, full remembering.

Some memories stay with us for life, while others stay for only a few weeks. When we are revising for exams, for instance, we demand a rapid consolidation of memory, within an hour or so, but those memories don't tend to stay with us for long. They may last a couple of weeks, or even a month or more before the exam, but most of them (the less meaningful ones) decay very quickly once the exam has taken place. More meaningful memories, like those to do with interpersonal relationships or information we really understand or use, last a lot longer. It is thought that this may be all about which of these areas is consolidating the memories. The consolidation done by the entorhinal, perirhinal and parahippocampal areas seems to be fairly rapid, taking perhaps just an hour, but those memories don't last as long. The consolidation undertaken by the hippocampus takes days or even months, but the memory is deeper and lasts longer.

Imagery and memory

Memories are all about how we store information, and when we're remembering things, we often use images of one kind or another. For example, we might store memories as pictures or images – perhaps of people we know in funny situations, or of scenes, or of actual pictures or cartoons that we have looked at. We can remember what some things sounded like. A familiar smell can bring back memories; and we store some memories as 'muscle memories' – **enactive representations** of what something feels like, such as going on a rollercoaster or taking a corner at high speed. And we remember what it is like to do things – to run, or to stretch, or to balance on one leg. All of these are forms of mental imagery.

There has been a great deal of research into how mental imagery works. Scanning studies have shown that using mental imagery activates areas right across the brain. It seems to begin with activity in the frontal and parietal regions, but also includes some areas in the temporal lobe. It used to be thought that mental imagery for any one particular sense would involve the same areas as dealing with external information of the same type – for example that visual imagery would activate the visual perception areas, or that auditory imagery would activate the auditory cortex. In a meta-analysis of over 60 different brain-scanning studies of mental imagery, though, McNorgan (2012) showed that it doesn't seem to be that simple. Although sensory imagery does involve areas very close to the primary perception areas for that sense, it often doesn't involve the main areas themselves.

McNorgan's research included studies of visual imagery, auditory imagery, tactile imagery, motor (movement) imagery, gustatory (taste) imagery and olfactory (smell) imagery. Each of these involved areas close to the primary cortex for that sense – for example, visual imagery involved areas in the occipital-temporal regions, while auditory imagery involved areas close to the olfactory cortex in the temporal lobe (see Figure 7.3). But there is also a general imagery network, which seems to be activated by all of these types of imagery. The areas involved in this network are scattered across the brain, but research has consistently shown that the top of the parietal lobes, the front part of the insula and the bottom of the left frontal lobe are all involved.

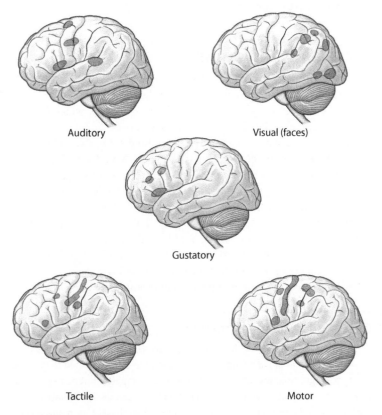

Figure 7.3 Imagery in the brain

It seems, then, that our memories of sensory experiences are processed in a slightly different way by the brain than the way we originally processed the experiences themselves. Psychologists studying memory have shown how flexible our memories are: they can be easily influenced by later events and often adjust themselves to be consistent with what we believe ought to have happened, or what we expected. We experience our memories as if they were factual recordings, but in reality they are nothing of the kind – as you'll discover if you watch a movie which used to be a favourite but which you haven't seen for many years. Your memory of some favourite bits is often slightly different from what you see in the film, because your brain has adjusted your memories. Our brains do this with imagery as well, not just with meanings, so the discovery that mental imagery activates areas right across the brain isn't all that surprising. What we remember draws on much more of our personal experience and uses more of the brain than the experience did when it first happened.

Forgetting and amnesia

All of this brings us to the question of forgetting – both the ordinary forgetting we do every day and the forgetting that comes about because of a problem with brain activity, which is known as amnesia. Forgetting is necessary in order for us to live our lives. If we always remembered everything we knew, we'd be unable to function cognitively at all. There would just be too much information. If we are to make decisions, select what we pay attention to, or have conversations with other people, we need to be able to remember what is relevant in that context and forget the rest for the time being. So context – both social and physical – is vitally important in memory, and we've seen how the brain codes it as an essential part of memory processing. That's why it is often helpful to recreate a context mentally when you are trying to remember something that won't come immediately to mind.

Case study: Too much memory

Forgetting is just as essential for a normal life as remembering. The Russian neurologist Aleksandr Luria studied a particular person who seemed to lack the ability to forget. Solomon Shereshevsky had been a reporter but he never took any notes, and eventually his editor, bemused by his uncannily accurate recall, sent him to Luria for examination. Luria studied Shereshevsky for many years and documented his experiences in a book, *The Mind of a Mnemonist* (1968).

While a young journalist, Shereshevsky's ability to recall everything had been a benefit to him, but as he grew older his mind became crowded with everything he remembered. He spent some years working in a travelling show, demonstrating amazing feats of memory, often using mnemonic techniques to ensure that lists were accurate and in the right order. But gradually he became haunted by old lists and information: a simple request would set off such a barrage of associations that he was unable to cope with it. He spent his later years in a mental haze; you can have too much memory as well as too little.

There's a common belief that people become more forgetful as they get older, but it's actually one of those popular myths. Some people do become forgetful as a result of some form of degenerative disorder like dementia, of course, but that isn't the case with most people. Studies that have compared younger people, in their twenties, with older people either just before or after retirement have shown that, in fact, older people experience less everyday forgetting than younger ones. But every time they do forget something – like going into a room and forgetting what they went in for – they notice it, and worry that they are becoming more forgetful because they are getting old. The younger people, on the other hand, don't worry about it, so they don't particularly notice the occasions when they are forgetful.

AMNESIA

Some forgetting, then, is perfectly normal, and nothing to write home about. But there are other forms of forgetting, which happen as a result of injury or damage to the brain, and these are known as **amnesia**. Amnesia can take several different forms, and much depends on the precise location of the damage. We have already seen how a taxi driver with damage to the hippocampus lost his detailed knowledge of the London streets but not his general knowledge of the layout of London and the main roads. Perhaps the most studied person in clinical history, Henry Molaison (referred to as H. M. in the clinical reports), was instrumental in showing how important the hippocampus is in memory.

In the 1950s Henry Molaison had an operation in which the hippocampus on both sides of his brain was destroyed in an attempt to control his severe epilepsy. The surgery resulted in a profound form of global amnesia, in which he was unable to remember the 11 or so years before his surgery (retrograde amnesia) and also completely unable to store new memories (anterograde amnesia). He was able to recall events in his childhood, up to age 16, but nothing that had happened since. Even several years later, he couldn't remember where he was living, who was looking after him, or even recognize himself in photographs – unless they were photos of his much younger self as a child.

One thing that emerged from these is that amnesia tends to affect what psychologists call declarative memory – our memory for events or episodes that have happened in our own lives, or for information that we have read or heard about in other ways. It doesn't usually, though, affect procedural memory, so people with amnesia are still able to use language, carry out everyday tasks and have conversations with other people. It also doesn't seem to affect short-term memory: H. M. was once able to retain a string of numbers in memory by going over it again and again for a whole quarter of an hour. But just a couple of minutes after stopping, he had no memory of even having tried to remember any numbers.

Case study: The hippocampus and memory

The hippocampus is one of the areas of the brain where new neurones are generated. As a result, people who experience mild damage to the hippocampus are often able to regain memory capacity, with sustained effort. In some surgical cases, though, like that of Henry Molaison, the removal of virtually the whole of the hippocampus is too much damage for neuroregeneration. Interestingly, though, although Molaison never recovered the ability to store memories consciously, he did either keep or regain some spatial memory. He had moved home to a bungalow in 1958. In 1966, neuroscientists asked him to draw its floor plan from memory, and he succeeded.

After H. M., many more people with amnesia and hippocampal damage of various kinds were studied. One woman with similar damage to that of H. M. showed some basic memory at another level: on one occasion, an experimenter pricked her with a pin while shaking hands, and the next day she was reluctant to shake hands. She didn't remember the event and couldn't explain her reluctance, but when she was asked to guess at a possible reason, she said that maybe people could hide pins in their hands. So she had some sort of memory, but no conscious awareness of it. The implication is that learning to avoid pain may involve neural processing in the spinal cord and brainstem and not involve the cortex or hippocampus at all. This would make sense, in evolutionary terms, as it's a very basic survival function.

As a result of this and other clinical evidence, researchers believe that it is the consolidation process that is affected when the hippocampus or its surrounding areas is damaged. The hippocampus and the medial temporal cortex seem to be the central route for establishing memories. We've seen how memories can be immediate or longer term, and there is debate about whether short-term memory is a first stage for long-term storage. But regardless of that, it is clear that memories do have to have a consolidation period before they become lasting, and the role the hippocampus plays in binding different aspects of memory into context may be essential for consolidating firm memories in the brain.

KORSAKOFF'S AND DEMENTIA

Not all amnesia relates directly to damage in the hippocampus. One common form of amnesia, **Korsakoff's syndrome**, results from damage to part of the thalamus itself, and is due to a chronic thiamine deficiency caused by excessive alcohol consumption unbalanced by nutrition – in other words, by drinking to excess without eating properly. People with Korsakoff's syndrome can hold conversations quite normally and may seem entirely unaffected in their everyday lives, until something happens which requires episodic memory, like what year it is, or where someone moved to, or who a significant public figure is. Their procedural memories carry them through the day, so their amnesia is not immediately obvious to the casual observer.

Another kind of amnesia can be the result of **dementia**. Dementia is a neurocognitive disorder that results in a gradual impairment of a person's ability to think and remember people and things; in emotional problems; and sometimes in problems with language, while not usually affecting consciousness. It can be caused by a number of brain diseases, the most common of which is Alzheimer's disease, affecting over half of people with dementia. Alzheimer's results from the development of plaques which form in the brain and push aside or inhibit nerve fibres, and also by tangles which develop among the neural fibres of the brain. It produces extreme shrinking of the cerebral cortex and the hippocampus, which means that the ventricles of the brain become enlarged as brain mass itself becomes less.

The extreme shrinking of the hippocampus is thought to account for the severe memory loss experienced by people suffering from the later stages of Alzheimer's, and the cortical shrinking is believed to account for some of the confusion and emotional disturbance experienced by sufferers. Alzheimer's is a progressive disease, with quite mild early symptoms but later symptoms being so extreme that an affected person may become a danger to themselves as they act out hallucinations or imagined scenarios, and need round-the-clock care. Perhaps the most distressing thing of all for relatives is the way that the disease can affect person recognition, so that in its later stages the affected person may not recognize even their own child or spouse, or may confuse them with another family member. This, too, appears to arise from the degeneration of those parts of the brain involved in person recognition.

While several theories have been put forward in an attempt to explain the origins of Alzheimer's, few of them have been conclusive. A few environmental factors may make people more vulnerable: smoking, pollution, synthetic food additives and the use of aluminium in cooking utensils have all been identified at different times as potential contributing factors. Some people have reported amelioration of symptoms as a result of removing such factors and keeping the person in a stress-free environment. But none of these factors has shown a conclusive connection, and research continues.

We've seen, then, how memory involves the frontal lobes of the brain, the hippocampus, and the regions around it.

Focus points

* Memory is complex and located in several areas of the brain. Working memory, which we use in thinking, involves activity in the frontal lobes of the cerebrum.

* Special areas of the brain deal with episodic and semantic memories. These areas, tucked underneath the cerebrum, allow us to identify familiar things and places.

* The hippocampus is essential for storing memories. It puts information into context, especially physical contexts, and this allows long-term memory storage.

* Memory often involves using mental imagery, and brain scans show that this mental imagery uses similar parts of the brain to those that the real sensory experiences would have done.

* While most forgetting is quite normal, amnesia can result from brain injury, disease or substance abuse. The form it takes depends on the type of brain damage involved.

Next step

Our memories are more than factual: in the next chapter, we will look at how the brain is involved in our emotional experiences – how we respond to rewards and pleasure as well as to negative emotions such as fear and anger.

8

Feeling emotions

In this chapter you will learn:

- ▶ *which areas of the brain are involved in reward mechanisms*
- ▶ *how emotions such as fear, anger, disgust and happiness arise in the brain*
- ▶ *where social emotions come from.*

Our emotions are an important part of what makes us human. As human beings, we experience a wide range of emotions: positive ones, like anticipation, joy, happiness and contentment, and negative ones like fear, anger and disgust. Emotion is the basis for all our fictional entertainment – video games, films, books and TV. But feeling emotion isn't unique to humans. As any pet owner will tell you, many animals share the capacity for emotional response. That isn't really surprising because, as we will see, a considerable amount of the way we process emotions in the brain concerns the subcortical structures of the brain rather than the 'higher' processing areas of the cerebrum.

Defining emotion isn't easy. What is the difference between a mood and an emotion, for instance? But there do seem to be some basics. Ekman (1992) identified eight fundamental emotions: happiness, sadness, disgust, anger, fear, embarrassment, awe and excitement. Each of these has different characteristic facial expressions and, Ekman argued, would have evolved because they had some kind of survival value. Ekman believed that the other emotions we experience were different facets of the same basic emotions; for example, he viewed satisfaction, contentment and relief as types of happiness.

Figure 8.1 Emotional facial expressions

Not everyone agrees with this classification, though. Other researchers have argued against the idea that there is a set of basic emotions, proposing instead that emotions are all about how we appraise situations – whether we judge them to be potentially threatening, pleasant or frightening, for example. Our emotional responses, in this model, depend on the more complex cognitive appraisals we make once the initial valence (positive or negative) has been established.

The real answer seems to be somewhere between the two. There do seem to be different brain responses involved in different emotions; disgust, for example, seems to activate different parts of the brain from anger, although some parts of its pathways are the same. We are a long way from identifying specific brain pathways for all of the 'basic' emotions. We do, though, understand quite a lot about how the brain responds to pleasant or rewarding stimuli, and that research goes back a long way.

Reward mechanisms

Back in 1954 the scientific establishment became very excited about a paper published by Olds and Milner, in which those researchers reported the discovery of a 'reward centre' in the brain. This was an area, found in human beings as well as animals, which gave every appearance of generating pleasure when it was stimulated electrically. Animals given the opportunity to receive stimulation to this area in response to performing a task, like pressing a lever, would do so repeatedly. People given a similar opportunity would also repeat the stimulation, describing it as intensely pleasurable. Experiments also showed that some animals would even ignore their basic needs for food or water in order to continue receiving direct brain stimulation of this kind.

Apart from providing plenty of material for science fiction writers then and since, this discovery generated a mass of scientific research. It focused on two areas: firstly, what areas of the brain were involved, and secondly, whether drugs could affect them and if so, what neurotransmitters were involved. Researchers quickly established that the same effect could be obtained from several areas within the brain, and that it seemed to tap into natural reward pathways that would normally have been stimulated by sensory activity of some kind. Within the brain, these reward pathways involved several areas in the limbic system, and also in the **diencephalon** – a general name given to the group of subcortical structures including the thalamus, hypothalamus, the pineal gland and part of the pituitary gland.

Investigations of the chemicals involved in these reward pathways showed that they tended to involve the dopamine and noradrenaline receptors in the structures concerned. Moreover, evidence showed that commonly abused drugs like amphetamine, cocaine, opioids and nicotine all tend to activate the same reward circuitry. Effectively, these drugs act on the neural receptors that would normally respond to either dopamine or noradrenaline, and so they stimulate the reward system. But researchers also found that both the electrical stimulation of 'reward centres' and the drugs imitating it had very short-term effects and gave no feeling of satiation – which was why animals receiving electrical stimulation would repeat their actions again and again. Unlike the natural reward feelings from, say, eating when we are hungry, they didn't get a message saying 'that's enough', so they didn't come to a natural stop.

Other studies showed that pleasurable anticipation, or looking forward to a reward, activates these pathways as well. In one study, single-cell recordings of dopamine neurones showed that those in the ventral striatum would respond to a signal that a reward was about to be given – in this case, that a monkey was to be given an treat of fruit juice. The same neurones were activated if the juice was given as an unexpected treat, but they were inhibited if an expected treat didn't materialize. Schultz et al. (1997), who carried out the study, proposed that these neurones are responding to the difference between reward and no reward – rather than to the reward itself.

Since that first discovery, brain research has developed many new methods, including brain-scanning techniques, and we now understand much more about how the brain works. The existence of reward pathways in the brain has been confirmed, although they are not exactly the same as the ones that researchers originally proposed. We know that information flows from many areas of the brain to three main areas that are involved in rewards. The **ventral striatum** (part of the basal ganglia in the limbic system) responds to rewards and also to the anticipation of rewards. The **orbitofrontal cortex** – that is, the part of the frontal cortex just above the eyes – integrates information from the ventral striatum with other inputs like emotional information from the amygdala, to calculate the motivational value of a prospective reward. And the top part of the **anterior cingulate cortex**, just above the corpus callosum which joins the two hemispheres, is involved in evaluating risks and working out whether an action is likely to be rewarded or punished (see Figure 8.2).

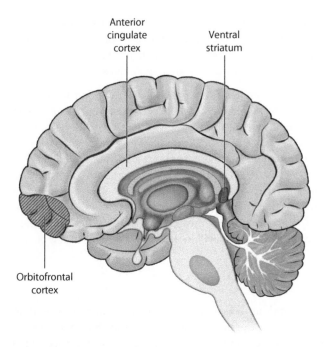

Figure 8.2 Areas on the reward pathways

In human beings, the orbitofrontal cortex is an important part of how we experience rewards. I've said that it calculates their motivational value, which is all about judging whether we think something will be pleasant or not. Eating a chocolate, for example, might normally be a positive reward, but if we had just scoffed a whole box of them it wouldn't work in quite the same way. The orbitofrontal cortex is an important part of the brain for keeping flexible: it assesses the internal (and external) context for the reward, and so judges how important, pleasant or rewarding the experience will actually be. People who have lesions in the orbitofrontal cortex don't adjust the value of rewards in the same way; their responses remain consistent regardless of their previous experience.

Blood and Zatorre (2001) used PET scans to identify the brain's responses to pleasurable music. They chose music that was intensely rewarding for their participants, selecting pieces described as particularly arousing: 'it sends shivers down my spine.' When they looked at the physiological and neural reactions involved at those times, they found activity in the areas of the brain known to be associated with reward pathways: the ventral striatum, the midbrain, the amygdala, the orbitofrontal cortex and the prefrontal cortex. The researchers pointed out that these are also the brain structures active in other types of reward, notably food, sex and some commonly abused drugs. So the rewarding experience we can get by listening to music we particularly like is similar to the positive experiences we get from other sources.

Reward, then, is an important part of how we experience emotions. But for most of us, our emotional experience is much more complex than that. Fear and anger, for example, are two of the negative emotions that come to mind. Feelings of disgust, both physical and social, are also very basic emotions involving special brain mechanisms, as are social emotions such as shame. Like many other aspects of psychological research, we know much more about the negative emotions than we do about the positive ones. There has been relatively less neurological research into happiness, for example, but there has been some. We will look at all these areas in this chapter.

Fear

The two emotions of fear and anger were the first of the emotions to be studied in any depth – mainly because they both produce such a distinctive aroused state in the body. Fear involves muscular tension, increased respiration and heartbeat and changes to the digestive system and bloodstream to release energy, among other things. You will find more detail about this in my book *Understand Psychology*, but what fear gives us is a physiological state known as the **fight or flight response,** so called because it helps an animal either fight an enemy or run away from it.

Fear responses are relatively easy to detect because they cause immediate changes in our levels of perspiration, making the surface of the skin less resistant to electricity. Because it's such a powerful emotion, fear responses are easily learned: snake and spider phobias are common, for example, and generally learned in childhood from other people's reactions. Children who grow up with people who are unafraid of snakes or spiders rarely develop these phobias. But the common nature of these fears means that it is relatively easy to use brain-scanning techniques to study how fear affects the brain.

One finding is that an area of the frontal lobes of the brain, known as the ventromedial prefrontal cortex, is particularly active in situations involving fear. It also becomes active when we are facing risky or threatening situations. This area of the brain has been shown to be particularly active in post-traumatic stress disorder (PTSD), and it has direct links with the part of the limbic system known as the amygdala.

The **amygdala** is a group of neurones buried deep at the front end of the temporal lobe – or, rather, two such groups of neurones because, of course, there is one on each side of the brain. It has connections with both the frontal and temporal lobes, and particularly strong ones with the areas of the temporal lobes involved in sensory processing and conceptual knowledge. It also has strong connections with the hippocampus and its surrounding areas, and we have already seen how important those are for memory. So it's not surprising that memory can play such an important part in fear reactions, and it is easy to see what evolutionary value this would have had.

The amygdala is actively involved in fear processing. It receives inputs from the sensory areas at quite an early stage – that is, from the thalamus rather than the visual or auditory cortex – and it also receives information from the hippocampus and the hypothalamus. Interestingly, the amygdala also has strong links with the olfactory processing areas. It's commonly said that animals can 'smell fear', and it has been shown that our own sensitivity to smells increases when we ourselves are in a fearful state. We saw in Chapter 5 how our sense of smell has links to very basic subcortical mechanisms, and the increase in perspiration produced by fear responses would be detectable by an animal with an acute sense of smell.

Case study: Emotion and the amygdala

The amygdala is strongly involved in how we perceive other people's emotions. One woman, known as D. R., experienced damage to the amygdala on both sides of her brain. As a result, she was unable to recognize facial expressions indicating fear and she also found it hard, though not impossible, to recognize anger and disgust. In fact, she had general problems with faces showing emotion: for example, although she could recognize famous people in different contexts, if they showed a different expression to indicate an emotion she didn't see them as the same person. Her brain deficit also left her unable to recognize emotions in speech, so it wasn't just about what she saw but about identifying emotions generally.

The amygdala can send messages to many areas of the brain: to the temporal and frontal lobes of the cerebrum, dealing with conscious awareness; to the hypothalamus, which acts as the regulator for bodily processes and has further links with the sympathetic nervous system which stimulates the fight or flight response; and to the ventral striatum, which forms part of a more general 'limbic circuit'. It's also involved in some forms of attachment, and we'll be looking at that more closely in Chapter 9. But its involvement in the fight or flight response and its strong connections with the hippocampus and sensory processing areas mean that it is inevitably connected with anger reactions, as well as with fear.

Anger

Interestingly, we don't understand nearly as much about the emotion of anger as we do about fear – mainly because anger is often paired with aggression. We know, of course, about the fight or flight response, which is activated in anger as well as fear, although not in quite the same way. And we know that the main role of the amygdala in anger seems to be in regulating aggressive responses so that they are less automatic and more appropriate to the situation. But most of our knowledge of the brain mechanisms of anger comes from studies of aggressive behaviour. The problem, though, is that aggression isn't necessarily the same thing as anger. People may act aggressively for many different reasons, and may not even feel particularly angry while they are doing so. We'll come back to the question of aggression in Chapter 12.

There have been relatively few studies of anger itself, but there have been some. Studies of people with lesions to the orbitofrontal cortex have shown that they are more likely to engage in reckless behaviour, and there is some suggestion that that they may become angry more easily than other people. It has also been shown that direct stimulation of the mid-brain can produce rage responses in cats, and that these reactions are also influenced by the amygdala and the hypothalamus. But how far these rage reactions have parallels with human experience is open to question, because stimulation of the same areas in other animals can produce sleepiness or relaxed behaviour.

Stanton et al. (2009) showed that increasing men's levels of the hormone testosterone, using injections, increased the neural activity of the prefrontal cortex when they were looking at pictures of angry faces. It also lowered the activity in the amygdala, which appears to control and regulate anger reactions. Pictures of neutral faces didn't produce this result, and neither did the same tests conducted with women. The hormone testosterone is important in development, stimulating particular aspects of growth and in maintaining the functioning of some neural circuits. Although it is generally thought of as a male hormone, women also produce testosterone, but men show much higher levels, particularly at puberty.

There is mirroring in the prefrontal cortex, as there is in many other parts of the brain. Researchers have found that when we look at other people's expressions, our neurones produce similar reactions to how it would be if we ourselves were having that experience. We'll be looking at this again in Chapter 9. But it does means that looking at angry faces produces a similar reaction in the brain to being angry ourselves, and this has been helpful in research into this area.

We can also re-experience our emotions when we remember them. Dougherty et al. (1999) used PET scans to study brain activity while people read accounts from their own lives of events that had made them angry. As they read them, they felt feelings of anger again, and the scans showed increased activity in the prefrontal cortex and the ventral striatum. These findings were also confirmed by fMRI studies. But the problem was that similar reactions were found in other emotional states, which means that activity in these areas isn't linked to anger as such, just with emotional excitement. However, studies using all three main types of scanning – PET scans, CT scans and fMRI scans – have also found consistent evidence that the ventromedial cortex is activated when people are reliving angry states.

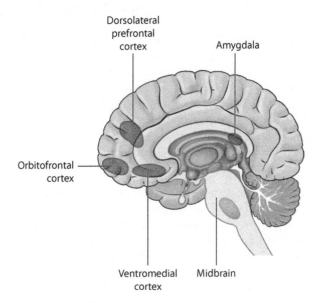

Figure 8.3 Anger in the brain

Overall, though, studies of anger as an emotional response to a situation are relatively few, certainly by comparison with studies of aggression; and most of them rely on asking people to relive their experiences of feeling angry, which may not be exactly the same thing. What we do know is that the amygdala and the prefrontal cortex are definitely involved, and that their activities seem to be able to override the inputs from the orbitofrontal cortex. Part of the prefrontal cortex, the dorsolateral prefrontal area, also becomes active when we are angry. As we saw in the previous chapter, that area is involved in working memory and its activation in angry states seems to inhibit the orbitofrontal cortex, perhaps reducing our overall level of cortical control over the way we react. This may be why we are so easily able to act irrationally when we are angry, and do or say things that we wouldn't do if we were thinking clearly.

Disgust

We don't often think of disgust when we think of emotions; we're more likely to think of happiness, anger or fear. But it is a powerful emotion, nonetheless, and one we can easily recognize in other people by their facial expression. Disgust is one of the universal emotions; that is, it is found in all human societies, which means we can assume that it is inherited as a reaction. But what we feel disgust about is partly learned. Things that some people accept as normal, other people may find disgusting – eating horsemeat or offal, for example. The word 'disgust' actually means 'bad taste', and the facial expression we make when we are disgusted looks as if we had tasted something bad, even when we are dealing with something entirely different. As an emotional response, disgust is likely to have evolved as a survival mechanism. If we feel an unpleasant emotion at the idea of handling rotten meat or human excrement, we are likely to avoid it or touch it as little as we can, which helps us to avoid being contaminated or catching a disease from the contact.

There is a particular part of the brain, known as the **insula**, which is directly involved in feelings of disgust. Not surprisingly, it is located very close to the primary gustatory cortex, which does the first cortical processing of the sense of taste. The insula, tucked right inside the cerebrum, has many connections with other areas of the brain such as the orbitofrontal cortex, the thalamus and the temporal and parietal lobes of the cerebrum. So it receives inputs from many areas, including areas to do with memory and social awareness. The front part of the insula also receives **interoceptive** information – that is, information about the internal state of the body, like how fast the heartbeat is or how tense the muscles are, and it is an important part of the brain for monitoring the bodily reactions produced by emotional states. We'll come back to some of the many function of the insula in future chapters, but it has a strong specific involvement in the emotion of disgust.

In fact, there are two major brain areas for disgust: the insula and an area of the brain known as the **anterior cingulate cortex**. The cingulate cortex is the part of the brain immediately around the corpus callosum, the band of nerve fibres connecting the two halves of the cerebrum; and the anterior cingulate cortex is the front part of it. It connects with those parts of the thalamus concerned with pain perception, with the orbitofrontal cortex, the amygdala and the insula itself. These other connections are involved in other emotions, too, but it is the insula and the anterior cingulate cortex that become particularly active when we are experiencing disgust.

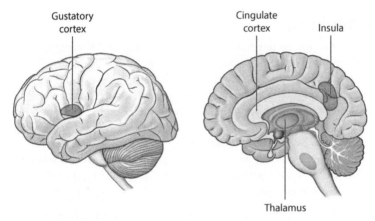

Figure 8.4 Disgust in the brain

Wicker et al. (2003) performed a series of fMRI studies of disgust, monitoring the brain's activity as people reacted to disgusting odours. As other researchers had also done, they found a strong response in the insula and also in the anterior cingulate cortex. But when they asked people to watch videos of the faces of other people being disgusted in this way, they found that seeing those other people experiencing the same emotion activated the same areas of their brain. There was a strong response in the insula and a less strong one in the anterior cingulate cortex. We saw in Chapter 6 how our internal representations of movement in the prefrontal cortex also mirror physical actions we see other people performing. The activity of the insula and anterior cingulate cortex shows that we have a similar mirror system in the emotion of disgust: we can share – at least partly – other people's experience by watching how they react.

There is another kind of disgust which we experience, known as moral disgust. This isn't triggered by physical or similar contaminants but by other people's actions, when we consider them to be both morally wrong and personally unpleasant to think about. They include things like acts of genocide, torture or cruelty. Interestingly, moral disgust appears to trigger the same brain mechanisms as physical disgust, involving the amygdala, the insula and so on. But it also involves more activity in the frontal lobes, particularly those involved in making judgements and personal perceptions. Using the same word to describe both the physical and the social emotion is therefore more than just a metaphor: we find both of them disgusting, and this is reflected in the brain's activity.

Happiness

Positive emotions like happiness have often been overlooked in the past, as if only negative emotions mattered. But we are gradually realizing that happiness, contentment, joy and other positive emotions are just as essential parts of our lives, and in many ways they are even more important. We know, for example, that positive emotions have a beneficial effect on our physical health. By allowing the body to escape the constant energy drain caused by the stress of negative emotions like anxiety, fear or anger, our physical resources are freed up to deal with other challenges, like tackling illness or fighting off infections.

A number of recent studies have explored the brain mechanisms involved in positive emotions, and one thing that comes through clearly is that they activate areas right across the brain. Suardi et al. (2016) found that when people are asked to remember happy events, the areas of the brain which are activated include those involved with other basic emotions like sadness or anger – the insula, the cingulate cortex and the prefrontal cortex. But they don't activate the same neurones. Instead, they set off subtly different groups of neurones in the same areas, and they also inhibit some existing pathways while exciting others.

Another area that has been found to be involved in the subjective experience of happiness is the part of the parietal lobe at the top end tucked down into the central fissure. It's part of the area of cortex known as the superior parietal lobule. MRI studies have shown that this part of the brain is active when we are feeling happy and that it also becomes active when our subjective experience is changing – for example, as we sink into sleep or as we calm down after being alerted by something. People who are generally happy and report more intense emotional experiences – negative as well as positive – have been shown to have more grey matter in this region of the brain than people who don't see themselves as particularly happy.

Scanning studies have shown a general tendency for positive emotions to produce more activity in the left hemisphere and relatively less on the right side of the brain. Several studies of positive emotions have shown increased activity in the prefrontal cortex of the left hemisphere but significantly reduced activity on the other side of the brain, in the right prefrontal cortex. This fits with the idea that negative emotions like fear and disgust, which trigger avoidance behaviour, are mainly processed in the right hemisphere, while positive emotions are processed in the left hemisphere. This is not clear-cut, though: both types of emotion activate both frontal lobes to some extent, and modern researchers are less inclined to make such a rigid distinction between the two hemispheres.

Case study: Anhedonia

There have been a few cases of people who find themselves unable to experience positive emotions – a condition known as anhedonia. In one documented example, Mr A. found himself in this condition after a drug overdose. Up to that point he had had a history of continuous drug use involving many different recreational drugs. When he recovered from the overdose, caused by swallowing all his methadone supplies so they would not be discovered, he had no positive emotional experiences but plenty of negative ones – being depressed, feeling hopeless and so on. He also lost his inclination to take more drugs. MRI scans showed that he had specific lesions to the globus pallidus – part of the basal ganglia – on both sides of his brain, but no other lesions or brain damage.

Another distinctive characteristic of positive emotions is that they seem to reduce neural activity in the area of the brain spanning the temporal and parietal lobes in both hemispheres. But neurones in the cingulate gyrus, the amygdala and the ventral striatum increase their activity with positive emotions, as they do when we experience other types of emotion. The areas are the same, but there is stronger activity in the cingulate cortex when people are remembering happy experiences in their lives than when they are remembering sad ones. The same is true for neural activity in the middle of the prefrontal cortex and at the top of the temporal lobe. Remembering happy experiences also produces significant activity in the cerebellum.

Of course, we can't be sure that remembering a happy experience involves exactly the same brain activity as being happy in the first place, but it's a little difficult to generate truly happy experiences for the first time in someone whose brain is being scanned! Nonetheless, it is possible to make three fairly definite statements about the neuropsychology of happiness. One is that positive emotions of this kind seem to engage more areas of the cortex than negative emotions do: their influence is broader right across the two hemispheres. The second is that many areas of the brain are involved in both happiness and negative emotions, but they don't involve exactly the same neurones. And the third is that there are specific areas of the brain that are activated by happiness and other positive emotions but not involved in negative emotions, including areas on the left prefrontal cortex, in the cerebellum and in specific parts of the ventral striatum.

Social emotions

Shame, embarrassment and guilt are some of what are known as the social emotions – those that are to do with seeing ourselves and our actions in relation to other people. Shame and embarrassment are all about how we believe others will perceive our actions. Shame is different from guilt, in that it necessarily involves other people's awareness or potential awareness, while guilt is about how we ourselves judge our actions – we can feel guilty even if other people could never know anything about what we have done. Interestingly, guilt is sometimes seen as a prosocial rather than an antisocial emotion, in that, having had an experience of guilt, people have been shown to be more likely to act in a positive way towards others. Ausubel (1955) regarded guilt as an important social mechanism: by providing a kind of 'brake' on antisocial impulses or behaviour, it helps people to adjust or maintain their interpersonal relationships, so they can get along in a reasonably harmonious, or at least an effective, way.

Embarrassment is often seen as a less extreme form of shame, and this is borne out by neurological studies of brain activity involved in the two emotions. Studies carried out using fMRI scanning show exactly the same types of brain activity when people were recalling shameful episodes as when they were recalling incidences in which they had been embarrassed. Apart from the more general areas of the brain involved in emotional processing, these scans showed heightened activity in the areas of the frontal lobe known as the medial and inferior frontal gyrus and in the areas of the temporal lobe known as the anterior cingulate cortex and the parahippocampal cortex.

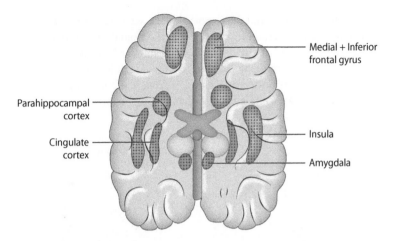

Figure 8.5 Shame and embarrassment in the brain

Guilt, on the other hand, produces a different pattern of activity in the brain. Michl et al. (2014) asked people to respond to sentences that had been shown to elicit feelings of either guilt or shame. The guilt-related stimuli also produced activity in the temporal lobe, but this was in the middle temporal gyrus and the fusiform gyrus; and they also showed activity in the amygdala and insula which was different from that produced by the shame-related stimuli. Similar findings have been obtained from other studies involving brain scans and these emotions.

There may be some distinction between activity in the left and right hemispheres as well. Takahashi et al. (2004) explored the differences in brain activity between shame and guilt, again using fMRI scans, and found that there was distinctly more activity in the right temporal lobe than the left one when processing embarrassing stimuli. But this difference didn't happen when people were feeling guilty. Takahashi used a Japanese sample, but the same thing was found in a similar study using Europeans, so we can be reasonably sure that the results weren't just a consequence of the different ways that shame and embarrassment are socialized in Japanese culture.

Social emotions in general have been shown to produce activity in the **orbitofrontal gyrus**, the part of the frontal lobe immediately above the centre of the eyes. They also involve activity in the neurones towards the back of the cingulate cortex, and in the front and top of the temporal lobes. And, of course, they produce activity in the subcortical structures involved in other types of emotional processing, such as the amygdala and the hypothalamus (see Figure 8.6). Overall, research into social emotions tells us that the frontal lobes are more associated with applying social norms and standards, while the temporal lobes are more associated with making judgements about one's own thoughts or behaviour, and making inferences about other people's minds. We'll come back to that idea in Chapter 9.

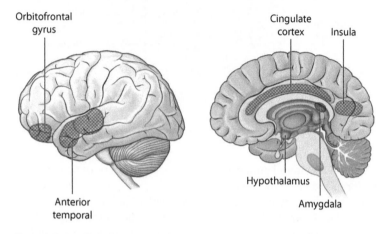

Figure 8.6 Social emotions in the brain

When we look more generally at how emotions are processed, then, we find activity right across the brain: in the frontal lobes, in the temporal lobes and in the limbic system. The exact pattern of activity varies slightly for different emotions: some involve more involvement of memory and experience, others of immediate sensory input. But the overall 'shape' of the neural activity is reasonably constant. Most of the neural pathways associated with emotions are general, regardless of which particular emotion is involved.

One part of the brain really stands out when we are trying to understand how emotion happens, and that is the amygdala. It is involved in both negative emotions like fear and positive emotions like happiness, and its connections range across the frontal and temporal lobes as well as with other parts of the limbic system. Through the amygdala, our emotional responses connect the primitive areas of the brain with the areas most concerned with memory and experience, and those to do with our decision-making.

Focus points

✳ The brain has a number of reward pathways, which include many subcortical areas and, in humans, the orbitofrontal cortex.

✳ The amygdala is active in all emotions, but particularly in fear and anger. Threat signals from the prefrontal cortex are also involved in these emotions.

✳ Disgust produces strong reactions in the insula and cingulate cortex, which are also mirrored when we see other people's expression of disgust.

✳ Positive emotions like happiness stimulate areas right across the brain, but with more activity in the right hemisphere.

✳ Social emotions activate the frontal lobes, the temporal lobes and the cingulate cortex, as well as other emotional areas like the insula and the amygdala.

Next step

The amygdala is also involved in our social experiences, which we'll look at more closely in the next chapter, in which we discuss relationships and how we form attachments.

9

Relationships

In this chapter you will learn:

- ► *about the areas of the brain that allow us to recognize faces and bodies*
- ► *how relationships and attachments use reward pathways in the brain*
- ► *which areas of the brain are involved in our experience of love and friendship*
- ► *how the pain of social exclusion and loneliness is similar to physical pain.*

We often hear it said that humans are social animals, and it's very true. Without other people, we would be unable to survive, or at least to achieve much more than mere subsistence living. But in social groups, we can support one another, develop and utilize our individual talents, and (eventually) build societies. Living in social groups, though, means we need to be able to recognize one another – to tell people apart and to respond appropriately based on what we know about the person concerned. In the modern world, we also need to be able to identify people whom we don't know personally. So recognizing faces is something our brains need to do – and in general, they are very good at it.

Recognizing faces

In Chapter 3 we looked at how we see objects. When we are looking at human faces, though, our brains react to them entirely differently from the way they look at other things. Faces are hugely important to us as social beings, and many of our responses to other people are based on our reactions to the facial expressions we see them making, and on the feedback we receive from how their faces react to our own actions or speech. So being able to recognize faces is an important part of social interaction.

PERCEIVING FACES

We have at least three brain areas that specialize in recognizing faces. The first of these areas is in the visual cortex – an area known as the **inferior occipital gyrus**. This part of the brain seems to be specialized to detect facial features, and it becomes active even when we are looking at stylized or cartoon-type faces. It's possible that it is the first face-recognition area in the brain to become active. Even from a very young age – just a few days old – babies smile in response to oval shapes with blobs arranged in face-like patterns, which is different from their responses to other shapes. As they grow older, the 'face' the baby responds to needs to be much more precise, until in the end they only respond to real faces or to images that are very similar. This innate sensitivity to face shapes tells us how fundamental other people are to us, as human beings. In adults, fMRI studies show that even distorted or schematic face shapes will activate the neurones in this area.

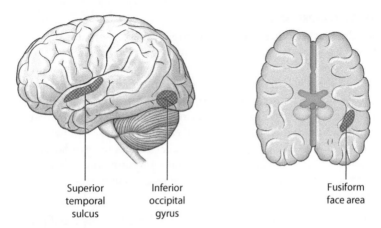

Superior
temporal
sulcus

Inferior
occipital
gyrus

Fusiform
face area

Figure 9.1 Face recognition areas

The second part of the brain that responds to faces is at the top of the temporal lobes, and is known as the **superior temporal sulcus**, or STS. It responds particularly to those aspects of the face that change during social interaction – things like lip movements, facial expressions and eye gaze. These are important signifiers of social meaning. We have already seen how the temporal lobes are actively involved in memory and meaning, so it makes sense that they would also be involved in understanding people's facial expressions. If we are asked to make judgements about what someone else is looking at, this is the area that becomes active.

If we are asked to identify a particular face that we are looking at, though, it is the third area that becomes active. This is the **fusiform face area**, or FFA, and it is located at the lower edge of the temporal lobes, tucked right underneath the cerebrum. It is very close to the memory areas we looked at in Chapter 7, and also quite close to the amygdala, which, as we've seen, is extremely active in our emotional reactions. The fusiform face area is the part of the brain that identifies the unique aspects of faces – those unchanging aspects of our faces that reflect our own personal identity. So we can see how important it is for our social interactions and why it has strong connections with our memory and emotional brain structures. People with damage to this area often suffer from prosopagnosia – that is, an inability to recognize faces – which can become a significant social problem.

READING EXPRESSIONS

Reading facial expressions is a different aspect of processing faces, but it is also an important part of everyday social interaction. It seems to involve two different routes in the brain. One of these is based on sensorimotor processing, and it draws on our brain's tendency to mirror what we see in other people. If we see another person with a distinctive facial expression – an expression of happiness, for example, or of disgust – studies have shown how our brains produce tiny changes in our own facial muscles. These changes reflect that emotion, providing what is known as facial feedback, which helps us to understand the emotion we are seeing.

Clinical evidence shows how people with damage to the facial parts of the sensorimotor area often have difficulty identifying emotions as well as expressing them. In 2008 Pitcher et al. performed studies that involved stimulating that area using transcortical magnetic stimulation (TMS). They found that the artificial stimulation affected how accurately people were able to read facial expressions, although it didn't affect their ability to recognize who those people actually were – their identities.

The way the brain makes sense of facial expressions is through a strong link between the initial perception of emotional expression, in the visual cortex, and the core areas of emotional processing of the brain – in particular the amygdala and the insula (see Chapter 8). This route becomes active if we are looking at familiar faces. We have an emotional response to people we know; it may be liking, love, dislike or something much more subtle, but it is enough to make a difference between how we react to familiar faces and how we respond to the faces of strangers.

There is a particular type of delusion, known as **Capgras syndrome**, in which people become convinced that family members have been replaced by strangers – body doubles who look identical but are not the same person. Some researchers believe that this happens because the neural pathways linking the emotional centres with the perception of that person have become disrupted or damaged. Without that emotional association, the person sees the individual and can recognize them, but still doesn't 'feel' that they are the right person. When we are looking at people we are close to, we expect to respond emotionally as well. Without that response, we would not be convinced of someone's true identity.

Case study: Facial recognition

Recognizing faces is a different brain activity from other types of recognition. This is vividly illustrated by two case studies of people who had localized damage to their fusiform face area. One man, R. M., was unable to recognize faces at all – even that of his wife. But he had a collection of over 5,000 miniature cars and he could recognize these easily. When he was shown a set of pictures of miniature cars, he was able to give the exact model, and when it was made, for 172 different examples. Another man had experienced a stroke that left him unable to identify faces previously familiar to him. After this event, he bought a small flock of sheep. They presented no problem: he could recognize each one of his 36 sheep, even if they were mixed in with others. Recognizing sheep was easy for him, recognizing faces impossible.

Recognizing bodies

It's not just faces that our brains are set to recognize so easily. We can also recognize people from their body shapes very quickly, even from quite a distance and sometimes just from the way they are moving. Two areas of the brain seem to become particularly active when we see human bodies. One of these is the **extrastriate body area**, an area just outside the visual cortex, which responds to generalized body outlines. It also responds to stick figures and other things that look quite like bodies, and it will even respond to body parts.

Other objects can also activate this area of the brain, but the more similar those objects are to human bodies or body parts the more strongly it responds. BOLD studies of brain activity in the extrastriate body area have shown that it produces very strong responses to human outlines and body parts like hands or legs, moderately strong reactions to images of animals – the more human-like, the stronger the response – and fairly weak responses to inanimate objects like chairs.

Studies of people with anorexia nervosa have shown that some of them have reduced amounts of grey matter in the extrastriate body area, and that this correlates with distortions in their perception of their own body size. If they are asked to compare their own body with silhouettes, they tend to see themselves as being much fatter than they really are. As with many brain findings, though, we don't know which came first – the disorder or the brain deficit. The reduction in grey matter could be a result of the problem rather than its cause.

The second area of the brain particularly concerned with body recognition is the **fusiform body area,** or FBA. It is located right next to the fusiform face area and partly overlaps it. Scanning studies show that this part of the brain seems to be concerned mainly with representing whole bodies rather than body parts. It is also active in representing body shape and size, but only in general terms rather than in making personal judgements about someone being fat or thin. The FBA also plays a significant role in person recognition: this part of the brain becomes particularly active when we are identifying a familiar person from a distance.

BODY MOVEMENT

Another cue we use if we are identifying people at a distance is the way they move, and this involves a third area of the brain, the superior temporal sulcus. As we've already seen, this area responds to the changeable aspects of faces. But it also responds to bodies, and in particular bodily movement and posture. Posture and movement are important signifiers of social meaning, so perhaps it is not remarkable that the same area deals with both types of stimulus. As we saw in Chapter 3, studies of bodily movement which involved attaching lights to key body joints and obscuring every other cue produces distinct activity in this area of the viewer's brain.

The superior temporal sulcus and the fusiform body area both have connections with the emotional areas of the brain. This is not really surprising, because body posture and movement can indicate emotional states as well as actions. Different types of posture and movement can indicate anger, fear, happiness, caution and many other emotional reactions; indeed, the word 'attitude' originally only meant posture of the body. It's because posture can be so revealing about our mental state that the word gradually changed its meaning. Evidence from brain-scanning studies has shown how emotional body language activates both of these areas of the brain. Interestingly, though, it also activates other areas, in the parietal and premotor cortex, which contain mirror neurones – groups of cells which reflect other people's actions in the same ways that they reflect our own. This mirroring is also an important part of how we interpret what emotional body movement actually means.

The way we make sense of faces and bodies, then, is an important part of social interaction and fundamental to our social lives. But as social beings, interacting with other people involves much more than simply being able to understand them. We also form relationships with other people, and our responses to those people that we are close to are qualitatively different from our responses to strangers or people we don't know well.

Relationships and attachment

Human relationships can be many and varied, ranging in feeling from passionate and intensely loving to gently affectionate or, in the other direction, ranging from feelings of mild dislike to active hatred. We experience an emotional response of some kind when we are dealing with almost anyone familiar to us; in some cases, that response forms a close bond between people. Anyone who has experienced the loss of a loved friend or family member will know how strong that bond can be; as with so many other things in life, we often don't comprehend how much certain people matter to us until they are gone, and the experience of bereavement can be profound and life-changing.

How does all this play out in the brain? It involves a combination of neural pathways and hormones, which interact in the emotions that we experience. Scanning studies of the emotional experience of maternal love, for example, have shown activity in the dopamine reward pathways of the brain and also in several other brain areas. We know, for example, that maternal love activates some specific areas of the basal ganglia, in particular the **globus pallidus** and the **substantia nigra**. It also activates the **Raphé nuclei** of the brainstem, which are known to be involved in serotonin reward pathways, and, of course, it activates those areas we have already encountered as involved in emotions generally: the thalamus, the insula and the cingulate cortex. This range of brain areas and pathways shows us how maternal love can be such a powerful emotion.

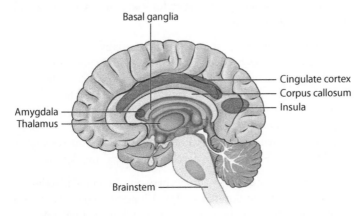

Figure 9.2 Maternal love in the brain

Not surprisingly, emotional attachments also involve the amygdala, the emotional centre of the brain, which, as we have seen, has connections right across the limbic system and the cerebrum. The degree of activation, though, varies according to the type of attachment: a number of studies have shown that activity in the amygdala is particularly strong in insecure attachments – those associated with some level of anxiety rather than more secure attachments in which people feel contented and safe.

This distinction between secure and insecure attachments was first made when researchers were investigating mother–infant relationships but, since then, it has been found to apply to adult relationships as well. It first became well established with Ainsworth's work with infants in 1978, in which she identified three types of attachment style. Some infants showed secure attachments, in that although they would become moderately upset if their mother went out of the room, they would soon adjust, and greet her positively when she came back. Insecure/anxious infants showed high levels of distress when their mother left and were difficult to comfort when she came back; while insecure/avoidant infants showed less disturbance when their mother left and tended to avoid contact with her when she came back.

Ainsworth found that these attachment styles were linked with the kind of parenting behaviour the child received (other researchers had established that it didn't only have to be the mother). Insecure/avoidant infants had parents who tended to leave the infant to its own devices a lot, while securely attached infants had parents who interacted consistently and appropriately with the baby, picking up on its signals and responding to them. Insecure/anxious infants experienced inconsistent parenting styles – sometimes attentive and loving and sometimes not.

Studies of how mothers react to hearing their baby crying do show some differences between mothers of securely and insecurely attached infants. In one fMRI study, Laurent and Ablow (2012) found that mothers of insecure/anxious infants showed more activity in the amygdala when they heard their baby crying than mothers of securely attached infants. Those with avoidant infants showed more activity in the prefrontal cortex, which the researchers believed was related to a higher level of emotional regulation.

There is some evidence that we may carry our attachment styles over to adulthood (but not inevitably). A longitudinal study showed a 72 per cent correlation between infant attachment style and attachments styles in adults. The general distinction between secure and insecure attachment is certainly one that characterizes the differences between many adult relationships: in some partnerships the participants are at ease with one another, showing a high level of confidence in their relationships. In others, the partners may be more anxious, worrying about how their partner feels about them and finding it hard to believe that their love is valid or real, which is typical of an insecure/anxious attachment. And some adults find it hard to trust other people and find it difficult to establish close personal relationships for that reason. These relationship styles tend to be measured using the two dimensions of anxious or avoidant relationships, with people who have secure attachments scoring low on both scales.

Some fMRI studies indicate that people with different attachment styles may use different types of mental control mechanisms in maintaining their relationships. Women with different attachment styles were scanned while imagining different relationship events, like arguments or break-ups. The researchers found that people with secure attachments tended to show activity in the orbitofrontal cortex, implying a more analytical approach to those events. Those with avoidant attachments, on the other hand, showed higher activity in the lateral prefrontal areas, implying a more personal and potentially emotional reaction. Putting it simply, people with secure attachments are less emotionally upset by arguments and more likely to look for solutions to the disagreement; while people with insecure attachments become distressed and upset about them.

Other studies have found a strong connection between attachment style and how the brain responds to images of happy, smiling faces. Scanning studies have shown how people with avoidant relationship styles tend to respond with low levels of activity in reward pathways when they see these images. People with anxious or secure attachments, on the other hand, tend to find images of smiling faces more rewarding, judging by the levels of response in the reward pathways, but those with anxious attachment styles also tend to show more activity in the amygdala. So our attachment style might influence even quite ordinary social interactions.

Love

It has long been established that there are many different types of love. Sternberg (1988) identified three components of loving: passion, intimacy and commitment, and showed how different types of loving, such as the passionate absorption of infatuation or the long-term affection of companionate love, can be distinguished as having different levels of these three components. Infatuation, for example, is high in passion but not in commitment or intimacy; while the companionate love between long-term couples is high in commitment and intimacy, but with less in the way of passion.

What is happening in the brain with these different types of love? Studies of intense romantic love have consistently shown a high level of activity in the **ventral tegmental area,** or VTA, which is in the midbrain, and also in the basal ganglia. Both of these are part of the dopamine reward pathway, which probably accounts for the experience of romantic love as a pleasant experience. But romantic love often contains an obsessive element as well, associated with the way that the person is often thinking of their partner with intrusive thoughts and images, even at times when they might be doing other things. This generates activity in several other areas of the brain – specifically the rear part of the cingulate gyrus, the septum, the caudate nuclei and the hippocampus.

Long-term love works a bit differently. Acevedo et al. (2012) conducted fMRI studies of a group of long-term happily married people, showing them images of their partner, a familiar acquaintance, a close long-term friend and a relative stranger. They found that looking at their partner's image produced distinctive activity in the dopamine and basal ganglia reward pathways which were very similar to those produced in studies of early-stage romantic love. But they also found that regions of the brain particularly associated with maternal attachment became active, in the limbic system and the midbrain, and including the thalamus and the insula.

Many of these areas of the brain involve the neuro-hormones oxytocin and vasopressin. They have been shown to be important chemicals in the formation of pair bonds in animals, and scan studies suggest that they may be equally important in the formation of long-term human partnerships. High levels of oxytocin are typically found in the blood plasma of people experiencing romantic love, but they are also found in people experiencing relationship worries, which may reflect the time they spend thinking about the relationship itself. Vasopressin is less well understood, but it may be involved in male sexual relationships, and it is also linked with reactions to stress.

Our deeply social natures mean that our relationships mean a great deal to us, so what happens when they are suddenly broken? As most people learn at some time in their lives, the experience of grieving is profound and incorporates a complexity of emotions, ranging from numbness and depression to confusion, guilt and even furious anger. These emotions, obviously, trigger the areas of the brain associated with those emotions. But grieving also generates activity in the areas of the brain involved with pain, particularly the anterior cingulate cortex and the insula. So the pain of grieving is a very real experience, and describing grief as 'painful' is more than just a metaphor.

For most people, the intensity of their grief does eventually subside. Even though they continue to miss the person and think about them, they learn to accept their loss, and adapt to their new life without the other person. But some people don't cope with their loss at all, retaining a morbid obsession with the person they have lost and refusing to adapt to life without them. This is known as complex grieving. Studies of these people have shown that they show more activity in the **nucleus accumbens**, an area at the back of the amygdala, than other bereaved people. This area is normally associated with reward and positive reinforcement, and it has been suggested that its activity may reflect a kind of condolence or dark comfort that the person obtains from yearning for the lost person and maintaining a continual sense of their loss.

Attachment and loving, then, are significant features of how we interact with other people, and they activate distinct areas of the brain. Although we are a long way from understanding everything clearly, we can see how loving is rewarding, and how grief is painful. These are not very surprising conclusions, you might think, but they do show why love can be such a strong motivation and how the pain we feel when we are grieving is not purely imaginary.

Friendship

Friendship and affiliation may not be as dramatic as the deeper forms of attachment, but they are still an important part of our social lives. Our everyday interactions with other people can be more important to us than we realize – so much so that, if they go wrong and someone behaves in an unexpectedly nasty or aggressive way towards us, it can upset us for the whole day, or even longer. Friendships play a crucial role in our mental health as well as in our social and psychological development. So how do they manifest themselves in the brain?

Gilman (2017) used fMRI to investigate how the brain responds to friendship. By asking people to respond to different types of relationship (for example personal friends or familiar celebrities) and relationships which generated different emotional reactions (liked, disliked or neutral), they found that four areas of the brain responded particularly strongly when people were interacting with their own friends. These were the amygdala, the hippocampus, the nucleus accumbens and the ventro-medial prefrontal cortex. The amygdala, as we've seen, is involved in positive and negative emotions; the hippocampus has to do with memory and person recognition; the nucleus accumbens is associated with reward and compensatory actions; and the prefrontal cortex is active in emotional regulation. All these areas are involved when we are interacting with our friends, but not necessarily when we are dealing with other people.

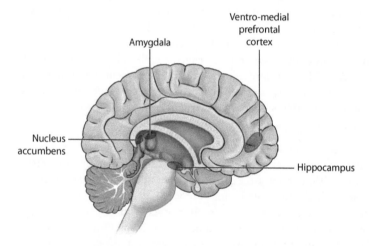

Figure 9.3 The brain and friendship

These areas are also involved in social activity, and peer influence can be a powerful factor in determining our behaviour. Gilman (2017) reviewed research into how the social influences on cannabis users affect the brain. We have known for a long time that social influences affect both the beginning of drug use and, in the case of cannabis, are often a strong factor in continued usage. Gilman's review showed that the brain regions involved in peer influence may operate differently in some cannabis users than among other groups of friends. In particular, the reward pathways of the brains of cannabis users showed more connections with frontal regions of the brain. Those reward pathways also became active in response to information about the rest of the peer group, implying that, for those cannabis users, information about their friends was more rewarding and involved more neural processing. This might imply that those people felt more dependent on their friendship group than the non-cannabis users did.

We now know that friendship activates many of the brain areas involved in loving and attachment, although often to a lesser degree. There are some other connections, too, which are not yet fully understood. Long-term friendship, for example, particularly activates the globus pallidus area of the basal ganglia, an area previously mainly associated with the control of movement. Its involvement in long-term friendship, and also in maternal love, is therefore slightly puzzling. Brain research progresses every day, though, and it might well be that by the time you read this all has been made clear!

Key idea

Although being a friend in real life is very different from being a 'friend' on Facebook, there does seem to be a general correlation between the two, in that people who have several real-life friends also tend to have many Facebook 'friends'. It might be a general measure of sociability, or what used to be called extraversion. But what is even more interesting is that a number of studies have found that people with more friends also tend to have more neurones in the amygdala, the part of the brain linked to emotional processing. But we have no way of knowing what causes what – whether their additional brain matter developed as a result of their sociability or whether they were more sociable because this part of their brain was larger.

Social exclusion and loneliness

It's a common metaphor to speak of social exclusion or separation as being 'painful'. But is it more than a metaphor? As we've seen, the experience of physical pain activates the front part of the cingulate cortex. So does experiencing the pain of social exclusion. Eisenberger, Lieberman and Williams took fMRI scans of people playing a virtual ball game, which was set up so that in one condition the person being scanned was included in the game, in another they were consistently left out of the game because of 'technical difficulties', and in the third they were left out because the other players preferred to play with one another and mostly ignored them. The researchers found that this social exclusion, unlike the other conditions, activated the same areas of the cingulate cortex that are activated by pain, and also activated the insula, which is also known to be involved in pain perception.

This shows that we experience the pain of social exclusion in much the same way as we experience physical pain. Other research, too, has shown that our sensitivity to social exclusion (as shown in fMRI scans) correlates with how sensitive we are to physical pain: the more sensitive we are to physical pain, the stronger the brain's 'pain' response to social exclusion.

Some people cope with social pain by turning to drugs, and it has long been known that the drugs that are most effective in suppressing physical pain are also used, or abused, to damp down mental pain as well. Opiates like heroin and morphine serve both purposes, and their use in damping down social distress was widely acknowledged in the nineteenth century. The pioneering naturalist Eugene Marais, for example, used morphine to deal with what he described as 'the pain of consciousness', which he believed was shared by apes as well as humans. Many drug treatment therapies focus on helping people to develop positive coping strategies to address their feelings of social exclusion.

Loneliness often has more to do with perceived social isolation than with actually being separated from other people. Some people find that they have become lonely because of circumstances or as a result of temporary shyness. They are often able to deal with it by joining new networks and making new friends. Other people, though, are chronically lonely: they still feel isolated and separate even when surrounded by other people. Because of the way they feel, they tend to look for signs of social rejection from others, and because they are so ready to see these signs, they often interpret other people's behaviour as rejection when it wasn't intended that way. They react by withdrawing further, which of course increases their loneliness.

In one fMRI study, chronically lonely people were shown images of various types of social or non-social scenes and their reactions were compared with people who were generally not lonely. The non-lonely people showed more activity in the ventral striatum (associated with pleasure) when they were looking at pleasant social scenes than when they looked at pleasant non-social scenes. For the chronically lonely people, though, it was the other way round. They showed more activity in the ventral striatum when they were looking at non-social scenes than when they looked at social ones.

When they were looking at unpleasant social situations, the non-lonely group also showed more activity in the area around the junction of the temporal and parietal lobes than the lonely people did. This part of the brain is concerned with analysing social information, remembering events and episodes, and other forms of mental analysis. The fact that it was activated in non-lonely people and not in lonely ones implies that the lonely people were simply accepting the unpleasant social scenes passively, while the others were taking a more active approach to analysing them, and working out why they were happening. They seemed to have more of a sense of agency in their social interactions, whereas the chronically lonely people saw themselves as passive receivers of social actions and unable to influence them.

There is also a similarity between the way the brain responds to physical warmth and the way it responds to social closeness and intimacy. Again, the language we use to describe relationships – for instance, describing people as warm or cold – reflects both our social responses and the brain activity associated with those physical conditions. Some researchers have suggested that this is because, as our social mechanisms evolved, they used the neural pathways and mechanisms associated with comfort and ease which had already been established earlier in our evolutionary history. As an interesting sidelight to this, it has been shown that chronically lonely people often take hotter baths or showers than non-lonely people do, which might be interpreted as using physical warmth to compensate for a lack of social warmth. Or it might just be that they prefer hotter showers.

Focus points

✱ The brain has several specialized areas that respond when we see faces or bodies. They connect with memory areas to allow us to identify other people.

✱ Attachment combines hormonal responses with brain activity, particularly in the reward pathways. Insecure attachments produce more activity in the amygdala than secure attachments.

✱ Romantic love is associated with areas in the dopamine reward pathway, but long-term love activates other areas of the brain as well.

✱ There are four main areas of the brain involved when we think about our friends. These are associated with person recognition, reward, emotions and emotional regulation.

✱ We experience the pain of loneliness and social exclusion in much the same way as we experience physical pain. Brain scans show that chronically lonely people may be more passive in social interactions than other people.

Next step

Another important part of being human and interacting with other people is how we use language. In the next chapter we'll look at how our brains make this possible, and what is actually involved when we are speaking with others and having conversations.

10

Communicating

In this chapter you will learn:

▶ *where the language areas are in the brain*
▶ *what happens in the brain when we listen to speech*
▶ *what brain areas are involved in speaking*
▶ *how problems with language arise.*

Language is one of the things that distinguishes us from the other animals that share our planet. While other animals have many different ways of communicating, and some can even learn to communicate with us, our use of language is unique. It extends to an immense degree our opportunities for learning, exploration, imagination and remembering.

Language is a social skill, of course. It originates with our need to communicate with others, and allows us to speak with other people, to hold conversations, and to share knowledge. But it also allows us to act as individuals: to store information, to modify what we do because we have learned from other people's experiences, and to plan for the future. We'll be looking at how the brain handles reading and writing, which are also important parts of our use of language, in the next chapter; but in this one we will look at the neural aspects of language and speaking.

Language areas in the brain

Language was one of the first human abilities to be identified as a localized brain function. The idea that certain psychological functions are located in corresponding parts of the brain goes back to the phrenologists of the nineteenth century, who attempted to read personality through the shape of the skull. This was in the belief that highly developed mental faculties came from highly developed areas of the brain, which would be larger than other areas and reflected in 'bumps' on the skull. The phrenological heads you sometimes see in antique shops (see Figure 10.1) were used to describe which areas did what, and the idea was widely accepted in Victorian times, even at times being considered acceptable as evidence in a court of law.

The popular belief in phrenology was reflected in many of the books of the time. In Charlotte Brontës novel *Jane Eyre* (1847), for example, Rochester asks Jane what she thinks of his looks:

> He lifted up the sable waves of hair which lay horizontally over his brow, and showed a solid enough mass of intellectual organs, but an abrupt deficiency where the suave sign of benevolence should have risen.

In response to her comment, his reply is:

> 'No, young lady, I am not a general philanthropist, but I bear a conscience;' and he pointed to the prominences which are said to indicate that faculty, and which, fortunately for him, were sufficiently conspicuous, giving indeed, a marked breadth to the upper part of his head...

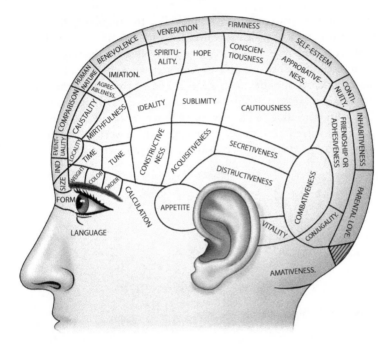

Figure 10.1 A phrenological head

Nowadays, of course, we realize that localization in the brain is nowhere near as simple. But nonetheless, there are distinct areas involved with particular functions. In 1861 the French physician Pierre Paul Broca identified the first of these. He had two patients with severe speech difficulties, in which they had problems producing speech but not with understanding what was said to them. When they died, Broca performed post-mortems on their brains and, in both cases, found a damaged area at the bottom of the left frontal lobe but no evidence of damage to the rest of the brain. He came to the conclusion that this must be the part of the brain that formulates words for speaking, and the area became known as **Broca's area**.

In 1874 the German physician Carl Wernicke identified another specific language area, this time at the top of the left temporal lobe. People with damage in this location had problems understanding what was said to them but could speak perfectly fluently themselves. Post-mortem examination of the brains of these people again showed localized damage, but this time to a specific area towards the back and at the top of the temporal lobe. This area became known as **Wernicke's area**.

Some other areas were identified, including the **angular gyrus** in the parietal lobe, which is concerned with identifying visual symbols of language, which makes it a key area for reading. But when researchers became able to scan active brains instead of dissecting dead ones or taking general EEGs, which were the main methods available to people before scanning, everything changed. They were now able to identify many other areas of the brain that are involved in our use of language, which were not apparent using the older methods. Let's look at these in more detail.

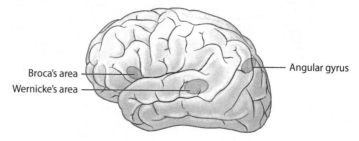

Figure 10.2 The language areas of the brain

LANGUAGE PATHWAYS

With most brain functions, as we have seen, we are dealing with pathways through the brain rather than single areas, and language is no different. Scanning research has shown that there is a **primary language pathway** through the cerebrum, which begins with Wernicke's area towards the back of the temporal lobe. This area receives information directly from the auditory cortex, in the case of spoken language, and the visual cortex in the case of reading or sign language, and it is mainly concerned with understanding – comprehending what the language actually means.

The second part of the primary language pathway is known as the **arcuate fasciculus**, which is a bundle of nerve fibres connecting the area at the back of the temporal cortex and the bottom of the parietal lobe with the frontal lobes of the cerebrum. It connects with other several areas on the way but, most importantly, it connects Wernicke's area with Broca's area. Broca's area forms the third part of the primary language pathway and, as we've seen, it is concerned with producing meaningful speech. This means that it, in turn, has connections with the motor and premotor areas at the back of the frontal lobe; we need to be able to move our lips and tongue in order to speak.

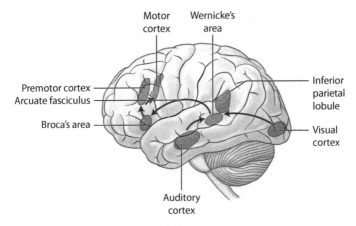

Figure 10.3 The primary language pathway

In the classic model, if we are having a conversation with someone, we listen to what they say, which means the information is processed first in the auditory cortex and also the visual cortex if we are looking at them while they say it. The information then passes to Wernicke's area, where we process it for meaning; and we respond to what we have understood by saying something ourselves, using Broca's area and those parts of the motor cortex which control the mouth and lungs.

Obviously, it isn't as simple and neat as that. If we look at people using language, PET scans of the brain show activity in many different regions depending on what is going on at the time. Listening to words, for example, activates both Wernicke's area and Broca's area and a large part of the left temporal lobe, producing activity right across the whole region. It also activates wider areas around the base of the frontal lobe, as well as parts of the motor and premotor cortex and some parts of the parietal lobe. In fact, when we're listening to someone speaking, almost half of the left cerebral hemisphere becomes active in some way – and a lot of the right hemisphere as well.

Speaking itself doesn't involve quite as much of the brain, but PET studies show that it activates both the premotor and sensory cortex as well as the motor cortex, as the brain plans the physical requirements of our words and converts them into instructions to the lips, tongue and voice box. It also involves some activity in a part of the brain known as the **inferior parietal lobule**, which is an area where the occipital and temporal lobes meet, near the back. Working out what we are actually going to say activates an even wider area: PET studies of people generating verbs show a great deal of activity towards the back of the frontal lobe and an equally large amount towards the back of the temporal lobe.

Listening to speech

As you see, there's quite a lot going on when human beings use language – and that's without even thinking about reading and writing, which, as I said, we'll look at in Chapter 11. Probably the best way to explore it is to look at what is actually involved when we are listening to speech and understanding it, before going on to look at the process of producing speech.

Perceiving speech begins when we hear someone say something. We saw in Chapter 4 that we process speech sounds differently from the way we process other sound information. Both of our cerebral hemispheres are activated when we hear someone speaking, but the language parts of it are processed more in the left hemisphere, while pitch and intonations are processed more in the right hemisphere. There is a special speech recognition area below the primary auditory area, and this is the place where the sounds are decoded and identified as potentially meaningful. The information then passes from there to the inferior parietal lobule, to be processed for understanding.

When we are listening to someone speak, we are often looking at them and lip-reading, which is why language processing draws on both visual and auditory information. That's fine while the two streams of information are in agreement, as they usually are, but there is an interesting illusion, known as the McGurk illusion, which happens when we receive different auditory and visual information. When people are presented with an auditory signal for one word, for example 'baba', dubbed on to a video of someone saying 'gaga', they don't hear one or the other. Instead, they perceive a totally different word – in this case 'dada'. The illusion is so strong that, even when people know it is happening, they still hear the constructed word. With their eyes closed, they hear 'baba'; with their eyes open and no audio, they see 'gaga'; but even after doing that, when they see both together they hear 'dada'.

What this shows us is just how strong both the visual and auditory pathways for language are. We can all lip-read to some extent: some people are better at it than others and some people train themselves to be really good, but we all have a basic competence in it. We might not realize it, but lip-reading can be an important part of ordinary conversation, particularly if the person we are listening to has an unfamiliar accent or is saying something technical. So while we might think that listening to speech is only about the sound, in everyday conversation it's actually about both sound and vision – and remember that we've evolved with face-to-face conversations. Auditory-only conversations, as on the telephone, are a very recent introduction to human experience.

There's another aspect to perceiving speech, which we have discovered only through the use of scanning. Brain scans show that when we listen to someone speaking, the auditory information we are receiving is also reflected by activity in the parts of the motor cortex that are to do with producing speech ourselves. Mirror neurones in the premotor cortex and in Broca's area become active, mapping what we see someone else as saying on to our own motor programmes for speaking. As we listen, we rehearse, mentally, the neural processes that would be involved if we ourselves were saying it.

Key idea

One theory of how we hear speech is that we unconsciously match what we hear to what we would do if we made the same sounds ourselves. This idea has been supported by the finding that there are mirror neurones in Broca's area and also in the prefrontal cortex, both of which respond when we hear people speaking. As we hear them speak, our brains unconsciously replicate the processing of saying the same thing. Interestingly, these mirror neurones also respond when we see people make gestures with meanings, like pointing, or shrugging and opening the hands to say 'I don't know'. Some people have suggested that this might indicate that language originally developed to amplify hand or arm gestures, but there could be other explanations.

Understanding speech

Understanding speech is quite a complex process. It begins with understanding words. That might seem obvious, but if you're listening to a foreign language, it's often quite hard to work out what the words actually are. As we speak, we tend to slur words together without realizing it – and if we do hear someone speak in such a way that they separate every single word, it sounds quite odd. Listening to a foreign language also sounds different from our own because each language has its own distinctive set of phonemes – the speech sounds that make up words. Infants babble all sorts of sounds, but as they learn to speak, they narrow that range down to the ones for their own particular language.

So the first step in understanding language is to decode the sounds we hear into phonemes, or language sounds. Then those phonemes have to be connected together in meaningful ways. There are many different theories about how the brain stores and accesses words, and many technical disputes. We're not sure quite how it happens, but we do know where in the brain much of it takes place, and that's in the area known as the inferior parietal lobule. It's quite a large region, where the temporal and occipital lobes form a kind of junction, and it lies in between the visual and auditory cortex and the somatosensory cortex. Wernicke's area is right next to it, joining on at the front. The inferior parietal lobule includes other areas that have also been identified as to do with language, such as the angular gyrus, which has long been associated with reading, and the **supramarginal gyrus**, which is associated with word choices and also, interestingly enough, with empathy.

This area – that is, the inferior parietal lobule – is important for processing information. It seems to be particularly concerned with labelling and classification, which is an important part of working out what things actually mean. It also has cells that can respond to several different kinds of information; for example, the same nerve cells might be activated by the sight of a ball, hearing the word ball, or feeling a ball in the hand. The area also has direct connections with both Broca's and Wernicke's areas. We have seen how those parts of the brain are involved in the primary language pathway, which includes the arcuate fasciculus, but the fact that they also have direct connections with the inferior parietal lobule means that there is also a second language pathway in the brain. It is possible for information to travel from Wernicke's area to Broca's area by a different route, supplementing the primary language pathway discussed earlier.

Wernicke's area, as we have seen, is directly involved with understanding language. It is the area that links together our memories and knowledge to make information meaningful. In order to do that, it has connections right across all of those parts of the brain. It also links with most of the areas to do with our knowledge of the world – the areas we looked at in Chapter 7, when we explored memory. But it uses more than just our memory of what words mean. When we listen to spoken language, we also take our knowledge of the person who is speaking into account: we might interpret a remark quite differently if it came from a friend than if it came from someone we disliked.

This involves the part of the brain that is particularly concerned with interpreting social cues and meaning, an area known as the **anterior cingulate cortex**. The cingulate cortex lies immediately above the corpus callosum, which is the band of nerve fibres joining the cerebral hemispheres. Anterior means 'to the front', so it is particularly connected to the frontal lobes of the cerebrum. Another area, just above it, is the **paracingulate cortex**, which is particularly involved in decoding and predicting social intentions. Both of these areas are important in understanding what people are saying to us.

The involvement of these parts of the brain means that we can also incorporate our knowledge of our culture. In some parts of the UK, for example, it is common to use 'deadpan' humour, in which things are said as if they were serious but there is an assumed common understanding that the speaker is being ironic. This type of humour often produces misunderstandings with people from other cultures; for example, people from the south-east of England often misunderstand humorous remarks made by people from northern regions by taking comments intended as ironic or sceptical completely literally. Wernicke's area receives information about words, categories and factual information, and incorporates it with knowledge of the speaker and the cultural context, to make sense of what is being said.

Most people in the world speak more than one language. It is not at all uncommon, for example, for people to speak one language at home and another when they are at work or socializing. Studies have shown that people who speak more than one language have more grey matter – that is, more interneurones – in their inferior parietal lobule than people who are monolingual. People who speak more than one language also seem to be more resistant to brain degeneration with ageing: they are less likely to suffer from Alzheimer's or other degenerative brain diseases, and if they do it is on average five years later than other people. Understanding and speaking more than one language appears to exercise the brain, allowing it to develop more grey matter and so have more reserves when it comes to ageing.

Speaking

When we are speaking, we begin with intentions – by deciding what we mean to communicate. This, as we've seen, involves the paracingulate cortex, which is concerned with our own intentions as well as those of other people. Then the brain needs to select the words we will need to achieve those intentions. To do that, we draw on our stored knowledge of vocabulary and on our knowledge of the person we are speaking to. We take into account what they are likely to understand as meaning, in the same way that we do when we are listening to someone speaking.

The brain also needs to work out the type of grammatical construction which what we want to say will need. Every language has its own type of grammar. The grammar we use in everyday speaking might not be the formal, precise kind found in a grammar textbook, but it still needs to conform to certain rules, which we learn very early in life. Even small children, for example, are able to recognize that Yoda in the *Star Wars* films used language that involved unusual grammatical constructions.

Even from an early age, then, the brain reacts to grammatical mistakes. We are very sensitive to what counts as appropriate grammar – so much so that a reliable change in brain activity occurs if we hear grammatical rules being broken. Studies involving EEG measurements have identified a classic spike in recordings – a sudden increase in the EEG response, known as an ERP, or **evoked response potential** – which happens when we come across grammatical errors. If we hear someone speak, or are reading something, and we come across something grammatically 'wrong' – such as 'It's wrong to shouted loud words' – our brain activity shows this distinctive spike. It's known as the P600 because it happens about 600 milliseconds after the beginning of the word concerned, and it concerns only grammar. If a sentence is nonsense but is still grammatical, the P600 ERP doesn't happen.

Incidentally, a similar ERP occurs when we come across a word that is wrong for the context. It doesn't happen at the same time as the P600, and it is concerned with meanings rather than grammar. It would happen, for example, if we heard the sentence 'The sky was full of white floating giraffes'. This response is called the N400, and it happens both when we hear a 'wrong' word and when we read one, which suggests that it has to do with processing language rather than hearing or seeing it.

These errors tell us that, when we are about to speak, the brain processes the grammar we are going to apply and the words we are going to use separately. It happens in Broca's area and the areas around it, but these areas also draw from our wider memories – not surprisingly. Once all that has been worked out, the brain then integrates it into in a plan for speech action. Like the other complex movements that we looked at in Chapter 6, it involves the prefrontal cortex in planning the overall action sequence, the premotor cortex in preparing the specific elements of the actions required, and the motor cortex in carrying out the movements.

Let's see how each of these areas might be involved in normal functioning. Imagine that you're with a friend, and you receive a message on your phone. You take longer than usual to read it. 'Who's that from?' your friend asks. 'Oh, it's from Jane,' you reply, scanning the email. 'She's moving house.'

If we look at the brain activity involved in this episode, the first thing is reading the message. This involves input through the visual system to the angular gyrus in the inferior parietal lobule, so that the information is interpreted as being language. Then it goes to Wernicke's area so you can comprehend its meaning. When you hear your friend's question, that information passes from your ears to the auditory cortex, on to the supra-marginal gyrus in the inferior parietal lobule for identification as speech, and then to Wernicke's area for meaning.

Your reply would draw information from the perirhinal area, which as we saw in Chapter 7 deals with memory for people and locations, and it would go via the arcuate fasciculus to Broca's area, where you would formulate the speech plans and words to express what you wanted to say. After that, it would go on to the prefrontal and premotor areas, and then to the motor cortex, which directs the muscular movements of your lips, tongue and larynx. This is a relatively simple model: on the way, of course, it would activate many outlying areas of the brain as well, but these would be the main areas involved. Yet we do it all completely automatically!

Key idea

Damage to the temporal lobes is often a consequence of diseases like Alzheimer's, and it can produce what is known as semantic dementia. People with this problem gradually become unable to understand individual words – usually the less common ones. They also find it more difficult to name objects that are different from typical examples of their category: for example, they could identify a starling from a picture because it's a fairly typical bird, but might find it harder to name a flamingo. They still retain their sense of grammar, though, and can put together meaningful sentences, but in the early stages they use simpler words than the ones they might otherwise have chosen. As the disorder progresses, they may begin to put apparently meaningless words into grammatical sentences – substituting them for the words they can no longer recall. These words are often not totally meaningless: they do relate to what the person is trying to say in some way, but they are not the right ones for the meaning they are trying to communicate.

Problems with language

Until the advent of scanning, brain researchers mainly had to rely on studying people with specific brain lesions or people with language problems, to infer what that could tell us about brain mechanisms. They were able to use forms of electrical monitoring as well, of course, and the error-detecting processes of the N400 and P600 were discovered in this way. Using these methods, researchers were able to learn a lot about language processing in the brain, mainly by looking at the different kinds of problem that some people experience when they are using language.

Problems with language are known as **aphasia**, and there are five main types: Broca's aphasia, Wernicke's aphasia, conduction aphasia, anomic aphasia and transcortical aphasia. Broca's aphasia, the first to be discovered, is mainly concerned with producing speech – that is, with speech output. People with Broca's aphasia don't usually have difficulty in understanding what has been said to them, or with reading. Rather, they tend to make mistakes when they are speaking or writing, and sometimes they find it hard to repeat words, or to name things, or just generally to talk fluently.

Some people with Broca's aphasia may choose the right words but get the sounds of parts of the word wrong, even when they know what the right pronunciation should be (we often pronounce words wrongly if we have come to know them only through reading, but that isn't a sign of aphasia). It's also known for some people with Broca's aphasia to communicate using what is known as telegraphic speech, which is limited to just the main words, without adjectives, conjunctions or the other 'extras' which convert a string of words into a grammatical utterance. As you might gather, then, there are many different types of Broca's aphasia: not everyone who is diagnosed with this problem will have the same set of problems. But they all involve difficulty in producing speech.

Wernicke's aphasia, on the other hand, isn't to do with producing speech as such; instead, it's concerned with understanding it. People with Wernicke's aphasia can speak easily enough, but they have problems in language comprehension – in understanding what has actually been said to them or what they have just read. In some cases, Wernicke's aphasia can affect a person's speech but this is generally by using words with the wrong meaning (like saying 'blue' instead of 'yellow'), or by a tendency to use nonsense words instead of real ones. For these people, the usual explanation is that there is some kind of deficit in the monitoring system that prevents them from checking what they are actually saying. People with this type of Wernicke's aphasia may also have problems repeating words and naming objects.

People who have the third type of aphasia, known as **conduction aphasia**, are generally able to speak normally, and they can also understand speech and read fairly well. They may show a slight impairment in speaking but otherwise seem fine, until they are asked to repeat what has been said to them. Conduction aphasia involves an inability to repeat spoken language, or to read aloud accurately. It isn't a problem in comprehending the language: they understand the meaning of the material perfectly well; it's just that they can't reproduce it. People with this language problem can show that they understand its meaning by paraphrasing it, so we know that it isn't about being unable to communicate the information either. Conduction aphasia is simply to do with replication – of either spoken or written words.

The most common sort of aphasia is probably **anomic aphasia**. This type of aphasia is all about difficulty in finding the right words for what the person wants to say. People who suffer from anomia generally have no difficulty understanding speech or reading and they can also speak normally. Their problem is finding the right words that they are looking for: identifying the right nouns, or naming objects. Anomic aphasia isn't about meaning: affected people can describe what it is they mean: 'You know, long-haired animal, lives in house, catches mice' quite easily. Unless the problem is very severe, they will also recognize the word if it is given to them. Sometimes, too, they can use the word they are looking for but in a different way: for example, being unable to remember the word 'comb' but describing it as 'a thing you comb your hair with'. Incidentally, most people experience a mild form of anomic aphasia from time to time – it can even vary in intensity at different times of day – without it being a problem. But for some it is an extreme problem, and one that can interfere considerably with their everyday communication.

The fifth kind of aphasia, known as **transcortical aphasia**, involves, as its name suggests, different areas across the cortex, interfering with most aspects of language functioning in some way. Each type of transcortical aphasia has its focus: motor transcortical aphasias mainly affect speaking, for example, while sensory transcortical aphasias are more concerned with the comprehension of language. But that isn't all they affect: they generally include a range of other problems in using language rather than just focusing on one area. What is interesting, though, is that they seem to be the exact opposite of conduction aphasias: they don't affect the ability to repeat language at all. People who have transcortical aphasia can read aloud accurately and they can also repeat parrot-fashion what they have just been told. What they can't do, though, is paraphrase it into different words or explain what it actually means.

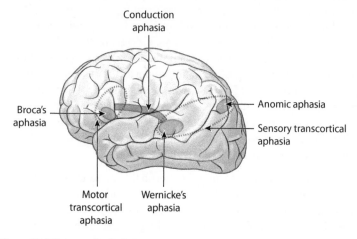

Figure 10.4 Sources of aphasia

Aphasias, then, can tell us quite a lot about how language is processed in the brain. Sometimes they are produced by specific brain damage, or lesions, in a specific area of the brain, and sometimes by an interruption to signal transmission, as might happen when brain cells are damaged by a stroke. In general, though, while our knowledge of the various kinds of aphasia has come from clinical evidence, scanning studies have confirmed the main areas that are damaged or impeded. They have shown, for example, how Broca's aphasia is associated with the anterior (frontal) language areas and Broca's area in particular; and how Wernicke's aphasia is associated with damage to the posterior language areas, and Wernicke's area in particular. Conduction aphasia seems to result from damage to the arcuate fasciculus, and anomic aphasia is associated with damage to the angular gyrus in the inferior parietal lobule. Motor transcortical aphasia is associated with general damage across the language centres in the frontal lobe, while sensory transcortical aphasia appears to be linked to damage to the language areas in the temporal lobes.

Brain scans also show us that this isn't the whole story. Other areas of the brain are also stimulated when we use language. But in this chapter we have looked at the main areas and gained an idea of how our brains process language, which is such a vital aspect of our social nature as human beings.

Focus points

* There are specific language areas in the brain and we also have language pathways linking different parts of the brain.
* Listening to speech draws information from vision as well as hearing. As we listen, motor neurones mirror the speech actions of the other person.
* Processing language for understanding involves understanding words, but also using social and personal knowledge about the speaker and the context.
* The brain uses both grammatical and word knowledge when we are speaking: EEGs show distinctive electrical responses when we come across inappropriate grammar or words.
* There are five different kinds of aphasia (speech problems), arising from damage to different parts of the brain. They are Broca's aphasia, Wernicke's aphasia, conduction aphasia, anomic aphasia and transcortical aphasia.

Next step

Speaking and listening are the most basic and early-evolved aspects of language use, but in the next chapter we will look at some more recent ones: reading and writing. And to make the set complete, we'll look at arithmetic as well.

11

The 'three Rs'

In this chapter you will learn:

▶ *how we acquire the ability to read*
▶ *about the different types of dyslexia*
▶ *how we learn to write and what causes agraphia*
▶ *what numeracy involves and what causes dyscalculia.*

Language may be what makes human beings special, but it is literacy that has formed the modern world. The widespread ability to read and write has made mass communication commonplace, limited the power of elites and autocrats, and is fuelling a major social revolution as we all become electronically interconnected. In this chapter, therefore, we are looking at the 'three Rs' – reading, writing and arithmetic. Yes, I know that two of them don't begin with an R, but describing them as the three Rs started out as a double-edged joke, from the way they sounded and as an ironic comment on bad spelling. By now, though, the saying has become so well known that it's just an ordinary figure of speech.

Reading, writing and arithmetic are all distinct human skills. Reading and writing are both sophisticated aspects of our use of language: they reflect and augment the considerable skills we acquire through our language ability. Arithmetic is related to reading and writing in that it is also about manipulating symbols, but it's different as well. While other animals can count to a greater or lesser degree, they are rarely able to perform anything but the most basic arithmetic, while almost all humans are able to perform quite complex manipulations of quantities – in real life through handling money, if not always in the abstract using paper and pencil. And the progression from arithmetic to the complex symbolic language of mathematics would be as far beyond the capacity of any other animal that we know of as writing a science fiction novel would be. How, exactly, does the human brain manage to achieve all this?

How do we read?

Reading presents an interesting challenge if we are trying to understand how the brain has evolved, because reading and writing emerged so late in our evolutionary history, and also because there are so many different alphabets and lexicons in the different human cultures. We couldn't have evolved reading in response to a basic, primeval survival need, as we did for abilities like movement, vision and the rest: primeval humans wouldn't have had any need for reading. Nonetheless, there are specialized areas of the brain that become active when we are reading, and which distinguish between letters, words and other shapes. How can this be?

The answer lies in the plasticity of the brain, which we discussed in Chapter 2. Although some parts of the brain have clear and definite functions, the brain also responds to the stimulation it receives. We know from the experiences of accident and stroke victims that we can retrain some parts of our brain to take on new tasks, with enough effort and persistence. And we have seen how neurones respond to learning by myelination, and also by developing more and stronger synaptic connections by preferring some routes to others.

In the case of reading, we find that there is a particular area of the brain on the rear undersurface of the cerebrum known as the **fusiform gyrus**. It's part of the general fusiform face area (see Chapter 9) but in a very specific location within it. On the right hemisphere, this area responds to faces – either real or as pictures or images; and the same applies to the left hemisphere in people who are not literate – that is, who have never learned to read. But in people who can read, this area of the left hemisphere responds to written words. As illiterate adults are trained to read, the left hemisphere version of this part of the brain becomes less sensitive to faces and more sensitive to words.

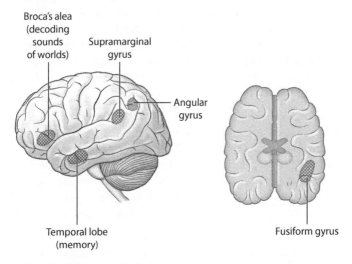

Figure 11.1 Reading areas in the brain

Given how important facial expression is in human communication, we might draw the conclusion that this particular area of the brain first evolved with the function of identifying meaning in facial expressions (which it continues to do in the right hemisphere of the brain). As human societies became more sophisticated, it then adapted to deal with meaning and communication indicated by other types of symbolism: symbols using sticks or images, hieroglyphs and so on. As written language developed and people went through the training and practice involved in learning to read, it adapted to interpret written words. What is also interesting is that the particular part of the fusiform face area that develops in this way is the part with the most direct neural connections to the language areas, implying that its stimulation is directly linked to communication using language.

Another area, right next to the fusiform gyrus, is known as the angular gyrus, and this is also important in reading. It receives information from the visual cortex and identifies specific shapes and forms as letters or symbols. So the angular gyrus identifies the visual input as letters or words and the fusiform gyrus interprets them for meaning, making sense of them in the same way as it does for facial expression in the right hemisphere. The area of the left hemisphere that combines the angular gyrus and the fusiform gyrus is often referred to as the visual word form area, or VWFA for short.

A third part of the general fusiform area on the left hemisphere, which is known to become active when we are reading, is known as the **supramarginal gyrus**. As we saw in the previous chapter, it is involved with language perception and also with empathy, and it becomes active when we are processing language for meaning. Working together, these three areas represent the levels of language processing involved in reading as a whole.

Incidentally, I should repeat the caveat I made in Chapter 2 here. We talk about the left hemisphere as the language hemisphere, to do with reading and writing as well as spoken language, and this is true for most people. But for some people, who are generally left-handed as well, the language areas are on the right side of the brain. Although we talk about the left hemisphere and language, it's important to remember that it isn't inevitably processed on that side.

These, then, are the main brain areas involved in reading. But reading itself is a complex process. The first step in learning to read, of course, is identifying the letters and words. This is specific to the particular language involved, and in alphabetic languages like English, Turkish and Russian it involves learning to recognize the symbols concerned, seeing how the symbols are combined to form words, and connecting the visual image of the word with the meaningful unit used in everyday language. Even recognizing the symbols can sometimes be tricky, since we often have different ways of writing the same letters, which is why texts designed for teaching reading tend to keep to one simple script.

This allows the learner to identify words by spelling out letters one at a time, and it helps while we are learning. Fluent reading, though, is quite different. Experienced readers identify whole words by their shape, and they adapt that recognition to the script they are reading. They don't need to stare at a word to identify it or break it down letter by letter; a glance is all that's needed. And in real fluent reading, this might not even be a direct glance. Studies of the eye movements made by fluent readers show that when they are reading whole paragraphs, they may fixate only once in a whole line of text. But as they are doing this, they take in the shapes of all the words either side of the point they are fixating on. They have become so skilled at reading that it just isn't necessary to look at every word. Getting to that level needs a lot of practice, though, to establish the relevant synaptic connections in the fusiform gyrus and the angular gyrus. Synapses in these areas, as you might expect, are more connected and well developed in fluent readers.

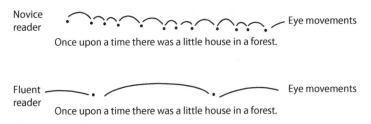

Figure 11.2 Eye movements in reading

READING IN CONTEXT

Reading in context also means that we are very sensitive to words that are incongruous or simply don't fit with what we are reading. In Chapter 10 we saw how researchers observing EEG patterns while people were listening found a distinctive negative spike of electrical activity: the N400. The same reaction happens when we read a word that doesn't fit with its context. Reading a sentence like 'The boat was out on the lake' wouldn't produce any particular reaction; but reading 'The boat was out on the tree' would produce a dip in EEG activity about 400 milliseconds after we began reading the word 'tree'. The N400 also happens when we encounter words that might appear plausible but conflict with our established world knowledge. A sentence like 'London buses are painted green' would generate the N400 in anyone familiar with the image of red London bus, but not in someone who wasn't. So studying the N400 shows us how the reading process draws on our world knowledge and expectations, as well as the information contained in the words themselves.

It can even show our unconscious biases. In a study published in 2017, Galli et al. reported how the N400 response to statements about Britain's membership of the EU, taken during the five weeks before the national referendum, was a more reliable predictor of how people were likely to vote than their conscious intentions. The researchers showed people various statements about the EU and Britain's involvement, both positive and negative. When they followed this up after the vote had been taken, they found that those who showed the N400 response to negative statements beforehand were more likely to have voted to remain in the EU, while those who had shown the N400 in response to positive statements had tended to vote to leave.

Reading, then, draws on our personal beliefs and preferences as well as our world knowledge and familiarity with the relevant symbols and words. What reading doesn't involve, though, is any connection with auditory inputs. Although readers often feel as though they are hearing the words they read spoken aloud, scans of the brain mechanisms involved in fluent reading show that they don't involve any connection with the temporal lobes, the area where sounds are processed. Fluent reading generates activity in the occipital lobes, where visual information is processed, and in the frontal lobes, where we process speech, but not in the temporal lobes. People who are learning to read sometimes show activity in the temporal lobe, as they try to interpret the visual word as being the same as the spoken word, but this disappears as they develop their expertise.

These findings first emerged from an early study by Posner et al. (1988), who used PET scans to explore what happens in the brain when people are undertaking different cognitive activities. First, they took a baseline reading of brain activity by asking their research participants to stare at a blank card. Then they gave them various cognitive activities to do, and compared their brain activity with that produced by the blank card. One of their findings, as we have seen, was that reading doesn't involve temporal lobe activity. But when they asked people to read pairs of words and think about whether they rhymed or not, they found considerable activity in the temporal lobe. When they asked people to read a word describing an object and then to think about a way to use that object, they found frontal lobe activity in both Broca's area and the prefrontal cortex, but no activity in the temporal lobe and not very much in the occipital lobe. Doing the same thing but from a heard word rather than from reading produced the same results, except for a slight activation of the temporal rather than the occipital lobe as the sound of the word, rather than the sight of it, was received.

All of this has implications for how we teach reading. While early methods emphasized the sounds of words, modern systems involve trying to encourage children to recognize words directly, using flashcards and other games. Other teaching methods emphasize speaking rather than listening: forming a direct connection between seeing a word and saying it out loud. Most methods for teaching reading, though, seem to be effective in the long run, as long as the child has enough practice. As with all acquired skills, it is only practice that makes it easy. Concern over the way that films and TV have replaced reading is particularly relevant here: for a small child, these forms of learning provide immediate results but the benefits of learning to read are less obvious. This is why reading stories to children, and showing them how reading can be a gateway into a satisfying imaginary world, can be just as important in helping the child to keep up the efforts needed to learn to read as any teaching technique.

Reading isn't just helpful to children, of course. Berns et al. (2013) explored what happens in the brain as adults become absorbed in reading – for example, in reading a novel. They conducted fMRI scans while people read thriller novels. While they were reading, these people showed increased activity in the left temporal cortex, and also in the somatosensory area. This remained for some time after they had finished reading the novel. Reading fiction helps people to imagine themselves in someone else's position, reflecting their actions and also seeing things from the point of view of someone else. It appears that this not only stimulates the imagination but also enhances empathy, exercising social and interpersonal processes in the brain in much the same way that visualization has been shown to enhance muscle memory in sports training.

Therapists have also found that reading, particularly reading positive and imaginative novels, can help people to cope with periods of intense stress. By immersing themselves in a different world, with less tension, they achieve a period of respite from their everyday worries and stimulate a different kind of brain activity. As Berns et al. showed, this can produce a lasting effect. It is possible that the therapeutic benefit happens because the stimulation of different brain areas helps the person to develop more constructive ways of thinking about their problems. Even if it doesn't, the temporary relief from everyday stress that they find while reading allows them a little period of rest.

Reading disorders

Some people, though, do have real problems with reading, which are different from simply not having had enough experience to do it fluently. One of the first types of reading disorder to be identified is known as **pure alexia**. This is a disorder in which people can spell out words letter by letter but are unable to recognize the whole word. It's called 'pure' because people with this disorder don't have problems with language or writing, or even with spelling: they can spell out words; they just can't recognize them. They can read, but only using a letter-by-letter reading method, which means that long words take more time to process and the whole thing takes much more time than it does in fluent readers.

Some researchers have taken this disorder as evidence that we develop a specific area in the visual cortex that recognizes words as such, which is separate and distinct from the area that recognizes letters. Pure alexia is generally seen as the result of interference with this part of the visual cortex, either through the interruption of its blood supply or because of damage to its connections.

Other forms of reading disorder are generally called **dyslexia** rather than alexia, because the person concerned can usually recognize words in some way, but not correctly. Dyslexia has been a controversial topic since it became fashionable to label anyone with a problem with spelling or reading 'dyslexic' (for more about this, see my book *Understand Psychology*). Nonetheless, some specific problems with reading do have a neurological cause. It may be surface dyslexia, in which people have problems with the appearances of words and letters, or deep dyslexia, in which people have problems with understanding words that are hard to visualize.

Key idea

The English language involves spelling that seems erratic, but it often tells us about the relationships between words. The silent 'g' in 'sign', for example, is there because of its relationship with the words 'signal' and 'significance'. But these complications mean that it is not at all uncommon for people to make spelling mistakes that have nothing to do with dyslexia. Frequent readers tend to have less trouble spelling because they are able to visualize words, having come across them so often. It is also noticeable that Chinese people learning to read English rarely make spelling mistakes: learning to read Chinese involves fixing the visual image in their minds, and they do the same as they learn to read English. A study of spelling mistakes in an entrance exam for prospective university students found that most of the mistakes seemed to come from

Research using fMRI scanning shows that people with dyslexia tend
to have interrupted processing in the area known as the inferior
parietal lobule – the area of the brain containing the angular gyrus
and the supramarginal gyrus. As we've seen, those are precisely
the areas that are involved when we are identifying letters and
interpreting words for meaning. It has been suggested that people
with surface dyslexia tend to have specific deficits in the angular
gyrus, while those with deep dyslexia tend to have deficits – that is,
problems that produce a lack of activity – in the supramarginal gyrus.

Weekes et al. (2016) explored the differences between Western and
Chinese languages in terms of reading and how words are processed.
In Western languages, for example, there are a relatively small
number of arbitrary symbols (letters) that can be combined in all
sorts of different ways to produce words. In Chinese languages the
symbols are ideograms, each of which represents a single morpheme,
or unit of meaning, and the tonal stress used to express them carries
a great deal of information about their meaning. There are many
more symbols involved, and the ways in which they are combined
and delivered is very different. Chinese people also use different
parts of their brain when they are reading, as shown in fMRI studies.
Weekes et al. found that this meant that reading disorders such as
dyslexia are quite different in Chinese from those in people using
Western languages. For example, in some forms of dyslexia, Chinese
people can recognize ideograms accurately but use the wrong tonal
stress to deliver them, which interferes with their perceived meaning.

Figure 11.3 Chinese ideograms

Writing and agraphia

The second 'R' of the three Rs is writing. Writing is another way of communicating using language, and it, too, draws on our knowledge of the symbols used by our particular language or cultural convention. But it's a complex process, combining as it does many of the cognitive aspects of language and reading with the need for fine motor control. In the past, handwriting was taught in schools as a specific skill, with considerable emphasis on the correct formation of letters and how they should be combined. This approach became less popular in primary education after the 1960s, so that although schools still teach handwriting to young children, there is less emphasis on developing precise scripts. It has remained an art form – calligraphy – but given the increased use of keyboards and texting for written communication, the emphasis on developing small, precise and attractive handwriting has largely disappeared. We still write, but perhaps not quite as carefully.

The brain mechanisms involved in writing are largely those we would expect. They include the areas concerned with reading: we have to know which letters and words we should be writing, so that includes the angular gyrus and the supramarginal gyrus. They also include the areas concerned with motor control and the planning and sequencing of actions, which as we saw in Chapter 6 includes the prefrontal, premotor and motor cortex. Interestingly, too, brain scanning has shown that writing also involves the language areas in the temporal cortex which are associated with the sounds of spoken words. It seems that we may rehearse the sounds of words as we write them, even though we don't do this when we read.

Problems in producing written language are generally known as **agraphia**. With so many different brain areas involved in writing, there are many different types of agraphia, and deficits in any of these areas can produce a difficulty. In general, though, we can divide agraphias into two main groups: central agraphias, which are also known as aphasic agraphia; and peripheral agraphias, which are also referred to as nonaphasic. These names give you the clue: aphasic is to do with language, so aphasic agraphia means that the person has some kind of problem with language which gives them difficulty in writing, while nonaphasic agraphia means that the difficulty comes from something else.

In one kind of central agraphia, known as deep agraphia, the person has serious difficulties with spelling. They are unable to remember how words look when they are spelled correctly (which is how

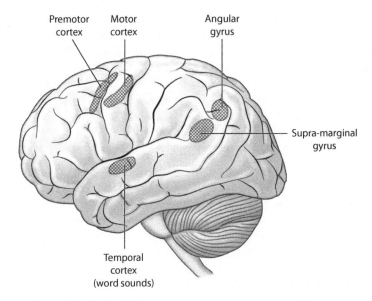

Premotor cortex   Motor cortex   Angular gyrus

Supra-marginal gyrus

Temporal cortex (word sounds)

Figure 11.4 Writing in the brain

many people remember spelling, particularly if they have read a great deal). And they are also unable to sound out words, to work out how they should be spelled. Perhaps not surprisingly, this form of agraphia is often associated with problems in reading and spoken language as well. Another form of central agraphia is known as lexical agraphia, and it is much more specific. People with this problem are not able to visualize what a word looks like, to spell it, but they can still sound it out. Their problem only becomes serious when they are having to write irregular words with odd spellings. People with a third type of central agraphia, phonological agraphia, have the opposite problem: they can remember what words look like but can't work out the spelling by sounding the word out. So, typically, they have more problems with the regularly spelled words than with irregular ones.

In peripheral, or nonaphasic agraphia, on the other hand, the person may have all the language skills they need for writing but has other problems that interfere with their ability to write, such as difficulty in executing the relevant motor actions. There are two main types of peripheral agraphia. Perhaps the more common of the two is known as apraxic agraphia, and it happens simply because the person has difficulty with the motor co-ordination needed to write the words. This might be caused by a disease interfering with motor co-ordination, like Parkinson's disease, or by partial paralysis arising from some other cause. Effectively, it is a problem of movement rather than of language understanding. The second most common form of peripheral agraphia is known as visuospatial agraphia, and this is caused by visual deficits that mean that the person doesn't write the words correctly, even though they may intend to do so. Problems like visual neglect, in which someone simply doesn't see one part of their visual field, are at the heart of this kind of agraphia. They are not to do with the language areas of the brain but from deficits in the visual system, which we looked at in Chapter 3.

Writing, then, is how we express language physically. It draws on most of the language processes identified in this chapter and in Chapter 10. As society progresses and more and more people use keyboards rather than handwriting, it will be interesting to see how the process of writing becomes redefined. Research into the neuropsychology of writing at present has tended to emphasize spelling and motor co-ordination; but as anyone who has ever tried to write an essay (or even an original message on a postcard) knows, there is much more to writing than that. The process of bringing together the world knowledge, language skill, learning and imagination that can be involved in writing is something we have yet to capture effectively in research. As we've seen, though, researchers have managed to explore something of the cognitive immersion involved in reading a novel, so it will be interesting to see whether the brain's activity when engaged in constructive or creative writing can also be studied in this way.

Numeracy and dyscalculia

The third 'R' is traditionally arithmetic. Arithmetic is a universal skill: in every human society we find that people are able to deal with quantities and numbers to some extent. People trade, they look after their animals, they count their possessions and so on. That's fairly straightforward and, perhaps not surprisingly, the human brain is well able to process numbers and quantities. Most of us can directly and pretty well instantly recognize quantities such as three and five, and some people can recognize larger numbers accurately as well. In fact, studies of even quite young infants have shown that they can discriminate pretty clearly between different numbers of dots.

This recognition appears to happen mainly in the parietal lobes, and involves two areas that are also involved in reading and language: the intraparietal sulcus and the inferior parietal lobule. It is thought that the inferior parietal lobule is concerned with simple, well-learned actions, like addition, while the intraparietal sulcus is involved in recognizing the numbers themselves. There are also neurones in the prefrontal cortex, next to Broca's area, which have been shown to respond to specific numbers, and other neurones in the same region which respond to amounts – generating stronger responses if there is more of whatever it is we are looking at.

We may be able to recognize numbers, but representing them is quite a different thing. All languages have some way of representing numbers symbolically, and different kinds of number symbols can be easier or harder to deal with. The Chinese numbering system is extremely straightforward, which might be one reason why high levels of numeracy are so much more common in China, while literacy is the skill regarded as more difficult – exactly the opposite of the English-speaking world. Processing numbers symbolically and doing calculations activates the two areas of the parietal areas already identified: the intraparietal sulcus and the inferior parietal lobule. People who have damage in these areas tend to develop problems in doing arithmetic or dealing with numbers, which is a disorder known as acalculia. We'll look at this a little later.

Numeracy, then, involves both the prefrontal cortex and the intraparietal sulcus of the parietal lobe. These are the areas that are mainly activated when we are dealing with numbers arithmetically, like doing addition or subtraction. It has been suggested that the prefrontal areas are the basic ones and that the parietal ones

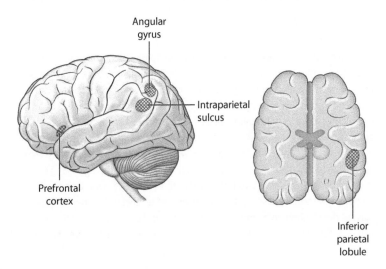

Figure 11.5 Numeracy in the brain

become involved later, as we learn more sophisticated uses of numbers. When Ischebeck et al. (2006) asked people to learn new multiplication tasks while they were being scanned, they found that it generated quite a lot of activity across the prefrontal region and slightly less activity in the intraparietal sulcus. When the participants were asked to perform similar tasks that they had already learned, though, it activated the inferior parietal lobule and also the angular gyrus – the areas we have come across in connection with language and reading. So as we become more proficient, we begin to process numbers using the same areas of the brain as we use in other symbolic tasks.

Split-brain studies have shown that we process numbers in both the left and right hemispheres, but that the left hemisphere is more exact, dealing with precise calculations, while the right hemisphere tends to deal with approximations. This ties in, too, with the way that language (which is also symbolic) tends to be mainly processed in the left hemisphere. People who are highly skilled at mathematics also tend to show a high level of literacy, but there's not such a strong correlation between the two abilities for most people. Most people are reasonably competent at manipulating numbers in basic arithmetic – in fact, much more so than they often realize, because they can manage money or real-world calculations easily. For true mathematicians, though, maths is like using another language – which is why people who are really good at maths are often not very good at teaching it! They are so fluent in it that they can't see why, or what, other people don't understand.

Key idea

Using symbols to represent numbers allows us to be much more precise in our calculations. One Amazon tribe has no words for numbers above 3 (which is used to mean 'many'), but they can perform all the complex calculations their lifestyle requires. For example, they can share a large group of things equally by putting things into different piles, and they can also compare the sizes of groups of things just as well as Westerners. What they can't do accurately, though, is exact sums. They are accurate with small numbers, but if they were asked to add, say, six and nine, their answer would be 'in the region of 15'. It might be out by one or two but no more than that, so they are perfectly numerate in terms of their cultural demands, as is reflected in their language. Having the symbols to express numbers allows us to do precise mathematics, which is needed in a modern society but of little value to a gatherer-hunter.

DYSCALCULIA

Some people are exactly the opposite: they have real problems when it comes to understanding or manipulating numbers, and this can be the result of a specific deficit in the brain. The disorder may be called either dyscalculia or acalculia, but as a general rule, the term **acalculia** is used if the person acquired their problem as a result of brain damage or injury, while **dyscalculia** is used if it seems to be a developmental problem, in much the same way as we use the term developmental dyslexia. Both acalculia and dyscalculia result in the person having difficulties in dealing with numbers in some way, but it isn't often a global problem – that is, a problem affecting all use of numbers. In many cases, for example, people who have dyscalculia can deal with 'real' numbers – like handling money – perfectly well, but they have problems when it comes to dealing with numbers as symbols.

This means that diagnosing dyscalculia is difficult, because we can also develop similar problems from faulty or misapplied mathematical learning, or from an emotional reaction to bad teaching. One of the most reliable ways of detecting dyscalculia is by scanning the brain's activity. Using fMRI, Dinkel et al. (2013) showed that dyscalculic children differ in the way that their brain reacts to simple calculations and number recognition. They show less activity in the visuoparietal cortex, which normally reacts to numerical information, but compensate for it in the frontoparietal cortex, particularly in the area relevant to the fingers. They may be counting on their fingers and doing simple calculations that way, instead of recognizing numbers symbolically.

Another difference between ordinary and dyscalculic individuals was found by a research team that used transcranial direct current stimulation (tDCS) at the back of the parietal cortex, in the area known to be involved in number processing. In the first study (Cohen Kadosh et al., 2010), they found that applying a positive current to the right lobe and a negative one to the lobe on the left hemisphere significantly improved the ways people tackled number tasks – and also that the improvement was still there six months later.

In a later study (Iuculano and Cohen Kadosh, 2014) members of the same research team tried the same method on two dyscalculic individuals, and found, interestingly enough, that although the same type of stimulation did produce positive effects, the polarity needed to be reversed, with the left hemisphere receiving the positive current and the right receiving the negative one. These were only preliminary case studies, and it remains to be seen how generally their findings can be applied. But the study suggests two things: firstly, that electrical stimulation may be able to improve arithmetical ability in dyscalculic as well as ordinary individuals; and secondly, that people with dyscalculia might actually have brains that respond differently to stimulation from people without it.

In this chapter, we've looked at how the brain processes reading, writing and arithmetic. There's much more to it, of course, and researchers are discovering new things all the time; but there's a limit to how much detail we can include in a book like this.

Focus points

* Reading involves identifying words in their context, so it draws on cultural and social knowledge as well as grammatical input. Scans show that reading stimulates the brain for some time after the activity itself.

* Deep dyslexia generally results from damage to the supramarginal gyrus, producing difficulty in understanding words, while surface dyslexia is associated with deficits in the angular gyrus, producing difficulty with word or letter appearances.

* Writing is a physical skill involving both motor and language areas of the brain. Central agraphia arises from language difficulties, while peripheral agraphia concerns motor co-ordination.

* Numeracy involves both the prefrontal cortex and the parietal lobe. We have specific neurones that respond to numbers as well as areas dealing with approximations and simple arithmetic.

* Children with dyscalculia have less activity at the back of the parietal lobe and more in the front of it than other children.

Next step

In the next chapter we will move away from the cognitive domain and get back to the social aspects of our lives, looking at how the brain is involved in how we understand ourselves and other people, and what it means to belong to social groups.

12

Us and them

In this chapter you will learn:

▶ *about the self and how we see ourselves*

▶ *about theory of mind and empathy*

▶ *how moral emotions involve brain areas involved in social processing*

▶ *how group membership can trigger emotional reactions*

▶ *how different types of aggression are linked to different areas of the brain.*

This chapter is about how we see ourselves and how we relate to other people. As human beings, we have a powerful tendency to sort our social worlds into 'us' and 'them' – to be aware of social groups and whether we belong to them or not. But that tendency is fluid, not fixed. We can belong to many social groups, and regard the same people as either one of 'us' or one of 'them', depending on the context. You might see your brother, for example, as one of 'us' when you're thinking about your family, but one of 'them' when you are thinking about males in general (if you are female) or when you are thinking about people who like, say, a particular genre of music.

How we see ourselves and others also has a strong effect on our feelings. Our brain mechanisms show how we can put ourselves in someone else's place, or empathize with someone, and also how we experience moral emotions – emotional reactions to how other people are behaving. Belonging to groups can make us feel pride or satisfaction, but it can also lead to aggression if two sets of people see themselves in competition. Modern brain science allows us to identify distinctive brain activity for all these experiences. But it all starts with how we see ourselves.

About the self

Who am I? We don't often ask ourselves that question: we usually think we know pretty well who we are and what we are like – although whether that knowledge is accurate or not is quite another matter! If we really think about it, though, the information that represents our own sense of self is quite complex and covers many different things. We might, for instance, think of ourselves in terms of our own bodies and physical capabilities – what we can do and what we feel we can't. Or we might think of ourselves in terms of the personal history that has gone to make us who we are – our memories, experiences and relationships. We might think of ourselves in terms of our ideas, motivations or goals – whether we are ambitious, energetic or simply accepting of what life brings us in a relaxed way. Alternatively, we might think of ourselves in terms of our friendship networks and social identifications – the nationality, profession and other social groups that we belong to.

Studying the self isn't straightforward, and many different aspects of how the brain works contribute to who we are. We have already seen, for example, how our motor and sensory systems give us feedback about our own actions and about the world in which we function. This information contributes strongly to our personal sense of our physical self and our sense of agency – that is, the feeling that we are able to interact effectively with the world around us. The social and cognitive skills we have developed also contribute to our sense of agency, allowing us to engage in complicated forms of interaction, and not just physical ones. We have seen how the brain deals with memories and emotions, and these too contribute massively to who we think we are.

While much of this book has already covered information about how the brain contributes to our sense of self, at some point all this information needs to come together to form our sense of identity. The region of the brain that seems to perform this function is known as the **medial prefrontal cortex**, or mPFC. It's located towards the centre of the frontal lobes, not at the very front but not very far in either, and it spans quite a wide area. When we think about ourselves, particularly in any kind of evaluative way, the medial prefrontal cortex becomes active. In one early study, people were asked to judge whether different personality traits were relevant either to themselves or to someone else. When they were thinking about themselves the medial prefrontal cortex became active, but when they were thinking of other people, areas of the left lateral prefrontal cortex became active – areas in this part of the brain that are particularly concerned with memory.

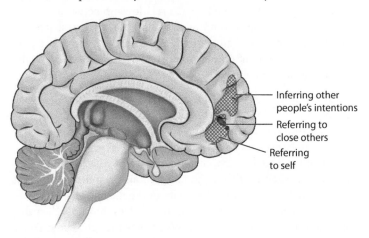

Figure 12.1 The medial prefrontal cortex

The medial prefrontal cortex becomes active if we hear someone saying our name. But it doesn't just become active when we are thinking about ourselves. It also activates when we think about people who are emotionally close to us – family members, spouses or dear friends. And it becomes active when we think about other things that are uniquely personal to ourselves. In one study, the medial prefrontal cortex was shown to become active when someone was looking at a photo that they had taken themselves, but not when they were looking at similar photos taken by other people. It can even respond to artificial representations of 'me'. Sui, Rotshtein and Humphreys (2013) showed people animated geometric shapes and asked them to think of one shape, such as a triangle, as representing themselves, with other shapes representing other people. They found that the medial prefrontal cortex became active when people were attending to their 'own' shape, but not when they were looking at scenes involving only the others.

There is some variation in exactly which parts of the medial prefrontal cortex become activated in different circumstances. For example, it becomes active when we are making judgements about other people relative to ourselves, but if the other person isn't particularly close to us, then it is the upper parts of the medial prefrontal cortex that become active. If we are making the same type of judgement about ourselves or people close to us, it is the lower regions of this area that respond. In other words, how the brain responds to judgements about other people varies in terms of how strongly we identify ourselves with them. Some researchers have suggested that we should see the relationship between ourselves and others as a continuum, rather than a hard, sharp distinction between 'self' and 'other'. Some people are so close to us personally that we think of them almost as part of ourselves, while others are more distant.

This varies quite a lot between individuals, of course, and it also varies from one culture to another. In 2007, Zhu et al. conducted a scanning study of Chinese and Western people. They showed them trait adjectives such as 'thoughtful', 'brave' or 'happy', and asked them to judge how appropriate these adjectives were to themselves, to their mothers, and to a famous person. As we might expect, the MRI scans showed activity in the medial prefrontal cortex whenever they thought about themselves, but not when they thought about the famous person. When they thought about their mothers, though, the Chinese people showed scan responses similar to themselves, indicating activation in the medial prefrontal cortex. This wasn't there with the Western group's responses when they thought about their mothers; those were more similar to the brain activity they showed when they were thinking about the famous person.

Some families are closer than others in all cultures, but in general this study highlights how some cultures, such as the Chinese, emphasize collectivism and interdependence, while others, such as American culture, emphasize individualism and independence. These differences are reflected in the sense of self that we develop (for more about this, see *Understand Psychology*). And brain research shows us that these differences are also reflected, at least partly, in the ways that our brains react.

Case study: Conjoined twins

Our sense of self is an important part of being human, and it can persist in even the most unusual circumstances. Tatiana and Krista are conjoined twins, whose skulls and a significant part of their brains are fused together. (Their brains were so closely fused that it wasn't possible to separate them surgically.) They feel each other's sensations: if one of them is tickled the other one reacts, and each of them can taste what is in the other one's mouth. Although they always speak of themselves as 'I', they nonetheless have distinct likes and dislikes: one of them likes ketchup, for instance, but the other one hates it. They share a great deal in terms of sensation and experience, but still have very different personalities, and despite sharing important parts of the brain, like having a single thalamus, they are definitely separate individuals.

The medial prefrontal cortex, then, is an important region of the brain when it comes to our sense of self. It brings together different aspects of our social and personal knowledge, and allows us to make inferences about ourselves and others. It helps us to understand feelings and intentions, which means that it plays an important role in how we understand other people. It also becomes active when we are recognizing irony or metaphors in the use of language; and it has been proposed as a key region for our theory of mind.

Theory of mind and empathy

Theory of mind, or TOM for short, refers to the ability to understand that someone else has a mind of their own and doesn't necessarily think in the same way as you. Having a theory of mind, and recognizing that other people will act according to the information that they have, is an important part of social living. We develop it as children when we are about three-and-a-half years old. Before that time, we are unable to predict what someone else will be thinking: we tend to assume that other people will think the same way as we do. Once TOM has developed, though, we are able to recognize that someone's different experience may give us a different understanding of a situation.

The classic study of theory of mind involves presenting children with a problem of this type: 'Sue hides a sweet in a box, while Tim watches. Then Tim goes out of the room, and Sue changes the hiding place of the sweet, putting it under a cushion.' The question is: 'Where will Tim look for the sweet when he comes back?' Small children tend to say that he will look under the cushion: they know it is there, so they assume that Tim will know as well. Older children, though, will say that Tim will look in the box: they recognize that Tim does not know abut the change and will act according to what he knows, rather than what they know.

Both scanning and lesion studies tell us that the medial prefrontal cortex is important for theory of mind. This is partly because it allows us to evaluate our own actions in relation to those of other people. For example, the medial prefrontal cortex becomes active if we are playing a computer game against another person (whether that person is present or not), but not if we are playing the same game against a computer. It's the belief that really matters, of course: if it was actually a computer but we believed it was a human, the medial prefrontal cortex would be activated; while if it was really a human but we believed it was a computer it would not.

One of the main functions of our theory of mind is that it helps us to predict what people are intending to do. Predicting social intentions is an important part of understanding other people, and it is fundamental to everyday social interaction. But predicting social intentions involves another brain area as well. This is the **anterior paracingulate cortex**. It's a layer of cortex right inside the brain and immediately above the cingulate cortex, which, you may remember, is a sort of band surrounding the corpus callosum. Anterior means that it is towards the front of the brain. When we are thinking about social intentions – including our own, but most especially other people's – this is the part of the brain that becomes immediately active: it, too, is involved in our theory of mind.

The anterior paracingulate cortex, then, becomes active when we are thinking about other people's intentions. The **posterior paracingulate cortex** – that is, the part of the paracingulate cortex towards the back of the brain – becomes active when we are thinking about ourselves, and how our own behaviour connects with that of other people. It is part of the complex of brain areas which are activated in the feeling we know as empathy – our ability to understand and share the feelings of other people. We have seen throughout this book how often mirror neurones become active when we see other people doing things. Our brains react as if we ourselves were doing that thing. There is mirroring in our motor systems, our perceptual systems, and even in our emotional systems, and these mirroring systems, at least partly, encourage us to perceive others in the same way as we see ourselves. That's an important basis for empathy.

Empathy is a complex experience, though, and it is thought to involve whole networks of neural activity rather than any one region of the brain. Scanning studies of emotional empathy have shown that it also involves the amygdala, which shouldn't be surprising given the fact that it is all about emotions and feelings. It involves the medial prefrontal cortex, too, and some other parts of the limbic system. The neural networks involving empathy also activate the area where the temporal and frontal lobes meet: on either side of the lateral fissure; the area immediately above the fissure, known as the inferior frontal lobe; and the area immediately below it, known as the superior temporal lobe. Part of this region, the **superior temporal sulcus** in the middle of the temporal lobe, is important in perspective-taking, and it's a significant contributor to our ability to detect and interpret social cues, like those indicating trustworthiness and intentions, as well as in empathy.

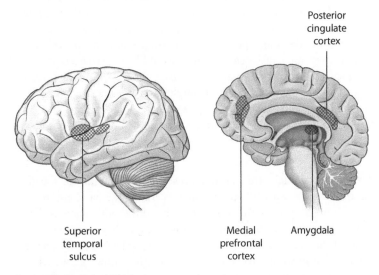

Figure 12.2 Empathy in the brain

We can see how empathy involves a range of brain areas, which in itself tells us something about the importance of how we react to other people – our social natures, in other words. These areas also overlap with the brain networks involved in theory of mind. Working together, they allow us to see things from the point of view of someone else, which helps us to engage confidently in everyday social action. Empathy isn't quite the same thing as theory of mind, of course, in that theory of mind is more about cognitions while empathy is more about feelings. But our ability to comprehend what others are feeling, as well as what they are intending to do, is an important part of social living.

Moral emotions

How we react to other people is also expressed in what are known as the moral emotions. The basic emotions are the ones we looked at in Chapter 8 – fear, anger, disgust and so on. The moral emotions, though, might equally well be called the social emotions, since they are all to do with either other people or the social aspects of our own selves. Some are prosocial, encouraging positive interactions with others, while others are more likely to produce an aversion, so that we will avoid the person or people concerned, or act negatively towards them.

There are a number of different moral emotions and several ways of classifying them. We can divide them into those which are all about how we think of ourselves, such as shame, embarrassment, guilt and pride, and those to do with how we think of other people, such as contempt, anger, compassion, gratitude and awe. We can also separate them into positive and negative emotions, in which case shame, embarrassment and guilt would be experienced as self-focused emotions that are negative or unpleasant, while pride is more positive. Awe and gratitude might be seen as positive other-oriented emotions, while contempt and anger might be seen as negative. Compassion or pity doesn't fall easily into either of those categories: it is perhaps more easily seen as an example of empathy and social awareness.

All of these, though, are emotional reactions to do with our social natures in some way. We saw in Chapter 8 how the basic emotions activate the amygdala and areas of the basal forebrain such as the insula. The same is true of moral emotions, but they involve other areas of the brain as well. If we feel that we, or someone else, have acted in a way we feel to be inappropriate for a situation, we will be drawing on our general social knowledge of events and how they ought to proceed. That knowledge has been shown to involve wide areas of the prefrontal cortex, so emotions like shame or indignation involve the prefrontal cortex as well as the amygdala and insula.

If we are responding to actions we feel to be morally wrong, on the other hand, we will be drawing on our knowledge of social concepts and principles, and this activates areas towards the front of the temporal lobes. Anger, embarrassment, contempt and guilt all involve those areas of the brain, as well as the insula and amygdala. And almost all of the moral emotions involve the areas of the brain concerned with how we perceive social cues. This, as we have just seen, includes the areas around the fissure between the temporal and parietal lobes, and it also involves other areas towards the back of the cerebrum.

The same areas are active in the positive moral emotions. Feeling gratitude towards someone involves using our social knowledge and our expectations about normal social procedures, as well as our feelings of personal relief or appreciation. But the reward pathways of the brain are involved in positive emotions, too, so feelings such as gratitude, awe and pride activate quite extensive areas of the cortex: social perception areas, the reward pathways, and any relevant sensory or memory areas. Incidentally, although being proud is often talked of as being negative, feeling pleased with ourselves and proud of an achievement is a justifiably positive emotion, which is why it activates the reward pathways that we looked at in Chapter 8.

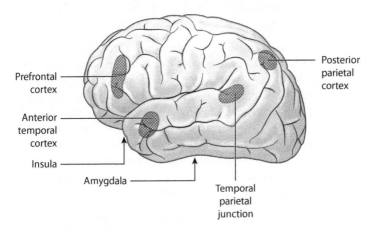

Figure 12.3 Moral emotions in the brain

Moral emotions, then, activate a range of brain areas. Some of those areas are the same as when we experience the basic emotions. Others include the areas of our brains that are involved in social processing. Moral disgust, for example, activates exactly the same areas of the brain as physical disgust: if we feel disgust at someone's behaviour, the same areas of the brain become active as when we feel disgust at rotten or mouldy foodstuffs. But it also involves the parts of the brain that concern our cultural and social perceptions of what is acceptable behaviour. What is considered to be disgusting can change over time, as society changes, or it can be different in different cultures; but the human brain can encompass a variety of social standards, and we pick up appropriate social knowledge and standards from an early age. Our experience of an emotion like moral disgust will be the same, but what triggers it may vary from one social situation to another and from one culture to another.

Group membership

Another thing that can trigger emotional reactions is when we meet, or think about, people who are very different from us. As we saw at the beginning of this chapter, our natural reaction, as social beings, is to see other people in terms of 'us' and 'them'. It's a process called social identification (discussed more deeply in *Understand Psychology*). We don't have just one category of 'us', though, since we all belong to several social groups and we identify with them in different ways. Any one person might identify with a particular job or profession, as a team supporter, as a member of a particular family, as male or female, and as being of a particular nationality – the list is endless! Try writing down your own set, and you'll soon see what I mean.

When we are engaging in a particular social identification, we adjust our thinking to fit it. For example, if I was thinking as a supporter of, say, Huddersfield Town FC, that would colour and shape the conversations and banter I had with a work colleague who, say, supported Leeds United. My colleague would become 'them', while other Huddersfield Town supporters would be 'us'. On another occasion, though, that colleague might belong to 'us' in our shared identity as employees of the same organization, for example if we were talking with people from a different organization, comparing working conditions or management styles.

Our social identifications, then, are important in shaping our social interactions. As part of that we have a natural tendency to categorize people into groups, classifying and sometimes even stereotyping them. In extreme circumstances, our reactions to 'different' people can become prejudice, where we judge all members of that group negatively. At its extreme, this can lead to discrimination, hatred and even genocide, so it is not surprising that it has been a matter of concern to many researchers, including neuroscientists.

What brain research has shown us is that stereotyping people isn't automatically the same as being prejudiced towards them. Reviewing a number of studies of race bias, Amodio (2009) showed that stereotyping is effectively a cognitive rather than an emotional process. People who were shown pictures of people from other ethnic groups but were not prejudiced showed increased activity at the back and to the left of the prefrontal cortex. They recognized the category and so were technically stereotyping, but they did not experience any particular negative feelings towards them. People who were racially prejudiced, on the other hand, showed increased amygdala activity, indicating an emotional reaction towards members of that particular ethnic group. Levels of prejudice were measured using an implicit attitude test, which showed up concealed negative attitudes even among people who claimed not to be prejudiced.

Amodio went on to look at research into the other neural processes involved in intergroup bias of one form or another. We have already seen how implicit stereotyping involves activity in the left posterior prefrontal cortex. This is the part of the brain concerned with concepts and memory retrieval, and also the priming and selection of information. But implicit bias that involves evaluation – in other words, that derives from prejudice – activates the amygdala, but not the prefrontal areas.

Experiencing implicit prejudice isn't the same as expressing it, either. We have specific areas of the brain that are involved in regulating social behaviour, which help us to respond appropriately to social cues. These areas are the middle part of the prefrontal cortex and the front part of the anterior cingulate cortex. Scans have shown how they become active if someone is implicitly prejudiced but not expressing it in a social situation. Someone may have grown up in a culture that encourages prejudice, for example, but have acquired very different values through their life experience. In that case, they would regulate their social behaviour and not express any stereotypes or implicit prejudice they might feel.

Cognitive control of this type involves different types of brain activity. Detecting bias or other social cues that indicate the need for social regulation of this kind produces activity at the back of the anterior cingulate cortex. Deliberately inhibiting stereotypes, on the other hand, produces activity in the right ventrolateral prefrontal cortex. And producing a deliberate, conscious response, regardless of whether it contains implicit stereotyping or bias, activates the anterior dorsolateral prefrontal cortex. So each of these types of control is different, depending on the social situation and on the perceptions of the individual.

This finding connects with the idea that stereotyping is largely cognitive while bias is largely emotional, or at least affective (to do with feelings). Amodio concluded from these studies that we need to adopt different strategies to deal with each. Challenging stereotyping, for example, might be best addressed by presenting the person with plenty of anti-stereotypical examples, encouraging them to learn to see the category in more complex and less simplistic ways. Challenging bias, on the other hand, would require more emotionally focused training, encouraging people to associate positive emotions and experiences with the other group, to counteract the negative emotions.

Aggression

As we've seen, stereotyping, or classifying people into groups, isn't necessarily the same thing as being prejudiced, and even those who experience implicit prejudice can learn that it is socially unacceptable to show it. In some cases, though, prejudice can become explicit aggression, and that is quite another matter. Aggressive behaviour towards members of other ethnic groups is illegal in many countries, and is rightly addressed by social sanctions of one kind or another. Unfortunately it still exists in some places, and there are other kinds of aggressive behaviour, too, which can be equally difficult to deal with.

Key idea

The enzyme monoamine oxidase is involved in the production of various neurotransmitters in the brain. Reducing the level of this enzyme seems to result in more general alertness and activity, so drugs that inhibited its production used to be used for chronically depressed people. However, reducing it can also produce higher levels of aggressive behaviour, and a genetic characteristic has been shown to do exactly that. Popularly known as the 'warrior gene', it's more likely to affect men than women. This gene has been around for 25 million years or so, and it has been suggested that it evolved in social primates because there was an advantage in having some males particularly prepared to defend the group. It's not carried by most of the population, though, because a social group in which everyone was highly aggressive just wouldn't survive. Men who carry this gene have been shown to become more strongly aroused in response to pictures of angry or fearful faces, with more neural activity in the amygdala.

Aggression has been defined as behaviour undertaken with the intention of causing harm. It's often considered to fall into two categories: **reactive aggression**, which arises from feelings of threat or frustration, and **instrumental aggression**, which is initiated by the person in order to achieve a particular goal. Bullying, for example, would be an example of instrumental aggression, whereas an aggressive act made in self-defence by someone who was being physically bullied would be an example of reactive aggression.

Within the brain, aggression is closely linked with the fear reaction. This is partly because both of these are involved in the fight or flight response – the reaction to threat shared by all mammals. The fight or flight response is a combination of various physical changes – increased heart rate, deeper breathing, release of adrenaline and so on – which all act to maximize survival for an animal faced with physical threat. There's more about this in *Understand Psychology*, but effectively what it does is release the body's stored energy, so the animal can either flee the threat, running away as fast as possible, or stand and fight for its life. In either case, there's no point holding anything back, so the brain triggers changes to release all the energy the body has available.

Humans have this reaction, too, even though most of the threats we encounter are not usually physical ones. This is why we react to threats either with fear and anxiety or with anger and aggression; and that's also why the brain activity involved in fear and aggression is so similar. We saw in Chapter 8 how important the amygdala is in the processing of fear, and it is equally important in aggression. There are different groups of cells, or nuclei, in the amygdala, and one group, the medial nucleus, has direct links with the hypothalamus. These links set off the fear or aggressive reactions in the body by stimulating the hypothalamus to pass the information on to the pituitary gland and other glands of the body, releasing hormones like adrenaline to keep the body's reaction going.

The hypothalamus also sends the information to a part of the mid-brain known as the **periacqueductal gray**. Direct stimulation of this area has been shown to produce rage reactions. But the periaqueductal gray also receives information from two other nuclei in the amygdala: the basal nucleus and the central nucleus, which 'switch' those reactions on and off. Signals from the basal nucleus of the amygdala to the periaqueductal gray appear to stimulate aggressive reactions, while signals from the central nucleus inhibit it. So the amygdala is actively involved in regulating aggression – that is, in making aggressive actions more or less likely in reaction to a threat.

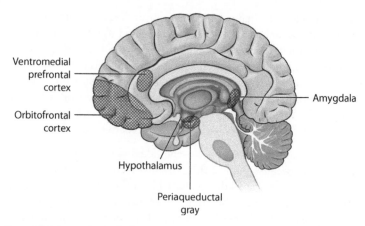

Figure 12.4 Aggression in the brain

There are, of course, other brain areas involved in aggression. We've seen how the frontal lobes of the brain are particularly concerned with the more complex aspects of our social lives. They, too, can be key in controlling or reducing impulsive behaviour. Two areas of the frontal lobes, the part known as the **orbitofrontal cortex** and the part known as the **ventromedial prefrontal cortex**, seem to be particularly important for this. In 2000 Pietrini et al. performed a series of PET studies in which people were asked to relive or imagine situations involving aggressive behaviour. They chose people who were not particularly aggressive but who had vivid visual imaginations. As those people imagined the situations, the researchers found much less activity in the ventromedial prefrontal cortex, implying that the ventromedial prefrontal cortex is actively involved in controlling and inhibiting aggression. Lesions to the orbitofrontal cortex of the brain have also been shown to produce an increase in aggressive behaviour.

Aggression, then, is only partly initiated by the cortex, at least in the case of reactive aggression, but it is definitely controlled by it. Both the frontal lobes and the amygdala are involved in controlling aggression. If you think about it, you'll see how these mechanisms would be essential for a species that depends on social interactions to survive. While some forms of aggression may be adaptive in some situations, for the most part, any form of society depends on its members being able to control their aggressive impulses. Our ability to be conscious of the consequences of our actions rather than driven by impulse, and to control how we act, is an active and well-developed part of how our brains work.

Next step

In the next and final chapter, we'll look at other aspects of what is involved in being conscious and thinking.

13

States of mind

In this chapter you will learn:

▶ *what influences how we make decisions*

▶ *how different types of consciousness are reflected in different patterns of brain activity*

▶ *how brain activity influences our sleep and dreaming*

▶ *how drugs can alter consciousness*

▶ *about the brain processes involved in social consciousness and humour.*

If we ask ourselves what we use our brains for, the obvious answer is that we use them to think with. But what is thinking? Philosophers and scientists have pondered this question for centuries, and have come up with various answers, ranging from the idea that thinking is just a random by-product of brain cell activity to the idea that it is thinking that defines our sense of being ourselves – as in Descartes's famous saying: 'I think, therefore I am'.

Thinking, as we can see if we stop to think about it, can take many different forms. There's reflecting on memories, noticing things we might not have perceived at the time. There's working out the answer to some puzzle or problem. There's thinking what to say in response to what someone else has said – either in conversation or in memory, thinking about what we should have said but didn't think of at the time. There's planning a project of some kind, whether it's a complicated meal, a craft project like making a model or building a shed, or working out what to say in a speech or a lecture. There's thinking about sport, as we watch it or participate, and as we weigh up the implications of the results. And there's thinking about decisions we need to make, working out what is involved and what is possible. All of these are examples of thinking, and I expect you can think of quite a lot more. All of this makes thinking very difficult to define.

In this book we've already looked at how the brain deals with many kinds of thinking: how it deals with memories, language, relationships, reading and sums, and interactions with other people. Thinking also includes being aware and conscious. We'll begin here by looking at decision-making: how we go about doing it and which parts of the brain are involved. Then we'll go on to look at consciousness: how it shows in the brain, what happens when we are asleep, and also the processes involved when we alter our consciousness by taking psychoactive drugs. And finally, we'll look at an aspect of social consciousness that shapes and colours our everyday social experience, lifting us from the strictly practical to the level of enjoyment and laughter.

Making decisions

Life is full of decisions to be made. Some of them are very simple and almost automatic, like deciding to take a sip of coffee. Others are much more complicated, like deciding to buy a house. What happens in the brain when we decide to take a sip of coffee is fairly clear. As we saw in Chapter 6, it begins with activity in the front part of the frontal lobe, where the decision is made. Neural activity then passes backwards to the prefrontal cortex of the frontal lobe, where the brain makes choices about what general procedures will be involved (observing where the cup is, whether it will need to be refilled and so on). These choices direct the next stage, further back in the premotor cortex, where the specific physical actions needed are planned, and that activates the motor cortex, right at the back of the frontal lobe, which tells your muscles to raise the cup to your lips and take a sip.

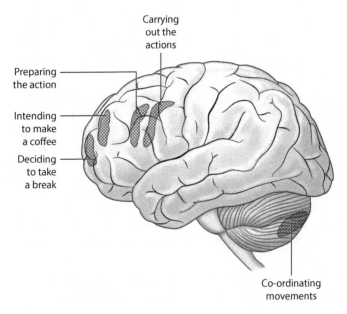

Figure 13.1 Sipping a coffee

Other types of decision-making, though, are a lot more complicated, and often not even rational. For many years, psychologists investigating decision-making assumed that it was a largely logical process. There were many examples of 'errors' in human thinking, but these were mainly seen as an otherwise logical system breaking down as a result of contextual or other influences. Gradually, though, psychologists came to realize that human decision-making is a far from logical process. Daniel Kahnemann summarized it in his book *Thinking, Fast and Slow* as having two distinct elements. He called one element **System 1 thinking**, which is intuitive, rapid and jumps to conclusions based on our previous experience and assumptions. The other is **System 2 thinking**, which is logical and painstaking, but much slower than System 1. On the other hand, if we are faced with a logical problem or decision, it is also much more accurate.

What Kahnemann and his colleagues showed was that, for the most part, we make decisions using System 1 thinking. The brain, he said, is lazy, preferring to fall back on its well-known strategies and assumptions, and that often leads to errors in logical problems. System 1 thinking has been shown to be vulnerable to many different kinds of bias. To give three examples: one is the availability heuristic, which is the tendency to choose the option we have seen most recently; another is the affect heuristic, in which our decision is biased by the emotions we associate with the various choices; and a third is own-age bias, whereby people are more likely to be influenced by the decisions of other people of their own age than by those of older or younger people.

There are more than 150 of these forms of bias, and any of them can influence our decisions. If we were deciding to buy a house, for example, our judgement might be swayed by which houses we have seen for sale most recently, by the types of houses friends of our age have tended to buy, and by our emotions and feelings when we view a particular house. Not many people will buy a house they have viewed immediately after an intense domestic argument, for example, because the emotions caused by that argument will still be affecting them and the house will become associated with negative emotions. And our decisions can also be influenced by any number of other biases.

What these biases do is tap into our existing social and personal knowledge. Scans have shown that they are reflected by neural activity in the relevant brain areas. The own-age bias, for example, shows neural activity in self-related areas of the cortex, while the affect heuristic shows activity in the amygdala and related areas. It's easy to condemn these biases as errors, and in modern social living those errors can be significant. But what is actually happening here is that our social experience and knowledge override logic and calculation, allowing us to respond rapidly and, in most social situations, reasonably effectively. Very little of our decision-making is purely logical or economically accurate but, in social terms, most of it makes sense given our own past experience and social situation. When it doesn't, as in the example of the argument, it is often tapping into older primeval brain mechanisms, telling us to avoid situations associated with pain or distress.

The areas of the brain that become particularly active when we are making decisions are mainly in the frontal lobes. The frontal lobe is our executive area: it controls important cognitive skills such as emotional expression, problem solving, judgement and memory. In decision-making, the part of the frontal lobe known as the **orbitofrontal cortex** becomes active, particularly if we are weighing up likely outcomes. Another area, the **medial prefrontal cortex**, becomes active when there is ambiguity or uncertainty involved in the choices we are trying to make. These two areas of the brain are almost always involved in decision-making.

There are other areas of the cortex involved in decision-making, too, depending on what type of decision we are making. As we saw in Chapter 8, the **ventromedial prefrontal cortex** (VMPC for short) is particularly concerned with processing risk and fear, and it's also active when we think about risks or have painful memories. So this part of the brain becomes active when we are weighing up potentially risky decisions, or making decisions that cause us anxiety. In the modern world there are many different types of risk, not just physical ones. But risky decisions, such as those involving money – deciding whether we can afford to take out a large mortgage on a house, seek a bank loan for a new car or pay the electricity bill on time – also activate the ventromedial prefrontal cortex, as we worry about what we can afford in the future.

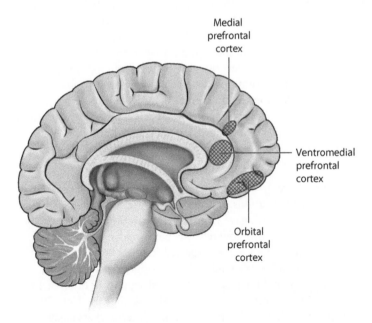

Figure 13.2 Complex decisions in the brain

Social or moral decisions involve different brain areas. In Chapter 8, we also saw how the **insula** is involved in negative emotions such as disgust and shame. The insula, a part of the cerebral cortex that is folded deep within the lateral fissure, has many other functions as well, including compassion and empathy, interpersonal experience and self-awareness. So it's not surprising that this area of the brain also becomes active when we are making social or moral decisions. The insula allows us to take into account our social knowledge as well as judgements about probability or risk, and this can make all the difference. When researchers used fMRI scans to see how people reacted to unfair decisions, they found that this area of the brain became strongly active – both when people experienced the unfair decisions and when they retaliated by acting unfairly themselves.

The overall patterns of activity in the brain can also tell us something about how we make decisions. There are distinctive patterns, or spikes, in EEG measures of brain activity, which are known as **event-related potentials** (ERPs). We saw in Chapter 10 how the ERP known as the P600 is associated with grammatical anomalies. There is another, known as the P300, which is associated with decision-making. It happens when our attention is drawn to something with a special meaning, different from the other stimuli around it. The P300 occurs right across the surface of the brain, and it's been shown to have two parts: one part is stronger over the frontal lobes and is associated with novelty, while the other is strongest over the parietal lobes and happens when we encounter things which are unlikely or improbable in that particular context. Encountering a Dalek, for example, wouldn't be improbable if we were at a science-fiction exhibition, but it would be if we were out for a walk in the country. An encounter like that would definitely catch our attention, producing a P300 reaction in the brain.

Researchers have used ERPs to examine how people react when faced with risky decisions. In a gambling task, Shuermann, Endras and Kathmann (2012) gave their participants the option of choosing a low-risk strategy with low rewards but not many losses, or a high-risk strategy with higher expected rewards but also higher losses. In general, people tended to prefer the low-risk strategy, but not always; and the researchers found significant differences in brain activity between the low- and high-risk choices. The P300 was associated with high-risk strategies, and the researchers suggested that this was probably because of the emotional aspect of high-risk decisions. High-risk choices also showed another distinctive spike in activity, known as the P200, which is associated with attention. That reaction also happened if people were given negative feedback about the choices they had made.

So decision-making involves quite a lot of the brain, but if we had to single out a particular area it would be the frontal lobes. They have been described as the 'control panel' of our personality, giving us the ability to make choices, to control our reactions and emotions, to project possible scenarios and action sequences, and to use our imaginations. But the frontal lobes don't act on their own: their strong links with the amygdala, the parietal lobe and other brain areas make sure that our thinking is informed by our emotions, our past experiences, and our stored knowledge. It all combines to produce our experience of thinking in general and decision-making in particular. We do make errors, but they are generally human errors, because we are not computers and we don't do strict logic very often.

Consciousness

Consciousness is another aspect of thinking. We see thinking as a conscious activity, but when we come to study consciousness itself, it turns out to be quite an elusive concept. Partly this is because our consciousness changes so much. We may be wide awake and alert, relaxed and dreamy, concentrating on something and focusing intently on it, performing everyday activities which involve some thought but can also be relaxing, like gardening or cooking, curiously pursuing an area of information, bored and fidgety; or any one of many other states of mind. Each of these involves a different kind of consciousness, and we experience many of them throughout each day.

Early studies of brain activity using electro-encephalograms (EEGs) were able to distinguish three different states of consciousness: normal waking, intense concentration and wakeful relaxation (see Figure 13.3). As we can see, normal waking activity shows variation in the amount of electrical activity (the amplitude) and also in its timing (the frequency). The frequency when we are concentrating hard is wider and our brain activity shows distinct rhythmical patterns, known as **theta rhythms**. When we are relaxed, the frequency of the brain's electrical activity is much smaller and the amplitude is also less, producing a small, tight pattern that also shows distinct rhythms. These rhythms are known as **alpha rhythms**.

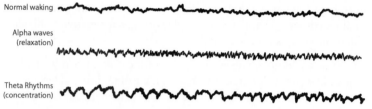

Normal waking

Alpha waves
(relaxation)

Theta Rhythms
(concentration)

Figure 13.3 Patterns of consciousness

We can learn to moderate our brain patterns and stimulate alpha wave activity using techniques such as mindfulness, or meditation. Both of these types of exercise involve learning to 'screen out' the distracting thoughts and images which are a normal part of everyday consciousness, to focus either on an inner awareness or on an awareness of our immediate surroundings, in order to achieve a relaxed and refreshing mental state. Zeidan et al. (2014) looked at how these work in the brain using MRI scanning. They found that three main areas generated increased activity when people were meditating or using mindfulness techniques: the ventromedial prefrontal cortex, the front part of the insula and the anterior cingulate cortex. Each of the areas they identified is involved in self-awareness in some way, and it appears that the focus on self and self-awareness is an important part of how these techniques work.

Those researchers also found that people using these methods experience an immediate reduction in general anxiety states. Because of this, they concluded that one of the reasons these techniques are so effective in helping people to live their day-to-day lives is because meditation or mindfulness allows people to regulate their own self-awareness, and in doing so, to control their anxious experience.

Case study: Neurofeedback

Alpha and theta waves are relatively easy to detect, and they are used in some 'mind-control' computer games. In these, a person (usually) puts a net of electrodes over their scalp, which reads the electrical activity of their brain at different points. Their task then is to move a pointer on a screen, or activate some other computer-generated image, by either concentrating or relaxing. With practice, some people can become quite adept at this. The biofeedback provided by the screen helps them learn how to moderate their brain's activity. More recent improvements in the sensitivity of the nets, particularly over the motor cortex, are beginning to produce games in which people can direct the movements of an avatar on screen, by imagining those movements themselves.

While we may be able to adjust our consciousness, what it actually is, in the brain, remains a puzzle. We have already seen some of the challenges presented by consciousness and the brain. In Chapter 3 we explored the phenomenon of blindsight, which happens when someone can react to a visual stimulus but is unaware that it has happened. A similar thing can happen with hearing, as we saw in Chapter 4. So we know, at least, that consciousness concerns the activity of the cerebral cortex, but that's not saying very much.

We may be approaching some understanding of it, though. Researchers have identified a part of the brain that does seem to play a powerful role in consciousness. This is the **claustrum**, a thin sheet of grey matter mostly composed of interneurones. It is located under the cerebral hemispheres, tucked in below the deepest part of the lateral fissure. It's only a couple of centimetres long, but it has connections right across the brain. Nobody is exactly sure what the claustrum does, but it forms links between the cerebral hemispheres, particularly with the areas involved in attention, and it also links with all the sensory areas and those concerned with the planning of movement.

Some researchers have described the claustrum as the seat of consciousness within the brain. The fact that it can bring together so many different modes of information suggests that it may be important in binding together our awareness, so that we don't experience information from the different senses as disconnected fragments but as a unitary whole. In one study, a woman who was given electrical stimulation to the claustrum became completely unconscious, with no awareness of her surroundings or reaction to stimuli. When the electrical stimulation stopped, she returned to normal but had no memory of the unconscious time. This has led to some suggestion that the claustrum might act as a kind of switch, turning our consciousness on and off.

Anaesthesia, sleep and dreaming

That still doesn't explain what consciousness actually is, however. Is it just about how much activity is shown by different areas of the brain? It's not really that simple. For one thing, it's difficult to say which mental state is more conscious than another, although we can (usually) tell the difference between being conscious and not being conscious – for example being rendered unconscious through sleep or through the use of anaesthetics. Even then, it can be tricky to define. Early studies of anaesthetics implied that consciousness was simply about the amount of activity shown by the brain. When they took PET scans of people during the process of becoming unconscious in response to a general anaesthetic, they found a gradual damping down of cortical activity right across the brain, until it resembled the quieter forms of sleep.

Studies using different anaesthetics gave different results. Some of them even seemed to produce a general increase in cortical activity as the person became unconscious. It turned out that those anaesthetics worked by blocking particular neurochemical pathways, specifically those involving the neurotransmitter glutamate. Blocking the action of glutamates in the brain produced an increase of general cerebral activity, but lower levels of consciousness. It seemed to dim awareness, while not stopping the actual activity of the brain.

What about sleeping? This is clearly a situation in which we lose consciousness, and brain researchers have studied sleep for many decades. When, back in the 1930s, they used EEG recordings to monitor the general activity of the brain, they found that there are several distinct levels of sleep that we pass through during the course of the night. In a typical night we cycle through these levels, going from the lightest level to the deepest level, level IV, and back again two or three times, and then, towards the later part of the night, going only down to level III or II.

It is worth noting that this is only one pattern, and there are other sleep patterns that seem to work just as well. Most people know that teenagers often sleep for much longer than other people, and this has been related to the amount of change going on in their bodies during this period. It has been suggested that the 'shut-down' time of sleeping may help the body to get on with growing and adjusting. Another fairly common pattern, particularly in mature adults, is to experience three or four cycles of very deep sleep in the first four hours or so, then a period of wakefulness for another couple of hours, and then another couple of hours of lighter sleep during the later part of the night. Because it's not as well known as the continuous pattern, some people worry that they have become insomniac, but those who accept the pattern and use their wakeful times to do things instead of lying and worrying about not sleeping, find that they can be just as refreshed by this sleep pattern as by sleeping continuously through the night.

Whichever pattern it follows, sleep still involves these four levels. They are named that way partly because of the pattern of the EEGs, but also because the numbers indicate how difficult it is for people to wake up from them: it's harder to rouse people from level

IV sleep than from level II. But interestingly, once they have been through a sleep cycle, people in what looks like level I sleep can be just as hard to rouse as someone in a deeper sleep. They also show rapid movements of the eyes, and when they do wake up they report dreaming. Because of these observations, this type of sleep is sometimes called **REM sleep** (REM for rapid eye movements) and sometimes paradoxical sleep, because it looks like a light dozing sleep but isn't.

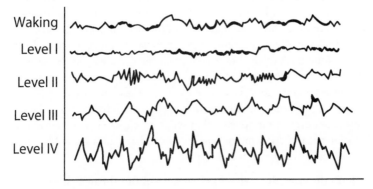

Figure 13.4 EEG patterns of sleep

Each level of sleep has its own distinctive EEG pattern, which is illustrated in Figure 13.4. As you can see, REM sleep has a pattern of rapid low-voltage activity, not all that different from the waking state, while non-REM sleep shows much more extreme highs and lows in its levels of electrical activity, with occasional 'spindles' of very rapid fluctuations. For a long time, researchers assumed that REM sleep was the only time dreaming occurred, but we now know that it takes place during other levels of sleep as well. People who wake from Level II sleep sometimes report dreaming, and sleepwalking and sleeptalking typically take place at levels III or IV. When people are woken during those episodes they also report dreaming. But there are also episodes of non-dreaming sleep during these other levels, and REM is a reliable indicator that dreaming is happening.

What is interesting is that dreaming doesn't produce a general increased level of activity across the cortex, as we might expect. Instead, it's associated with what Siclari et al. (2017) referred to as a 'hot zone' – an area of the brain that spans the parietal and occipital lobes. In this area, Siclari and her colleagues found a reliable decreased pattern of low-frequency activity associated with dreaming, no matter whether the person was in REM or NREM (non-REM) sleep. The researchers argued that this pattern of decreased low-frequency activity is an even more reliable indicator of whether someone will report dreaming or not than whether they are showing rapid eye movements.

The researchers also found that the dreaming experience often involves bursts of high-frequency activity in different parts of the brain. When they woke their participants and asked them what they were dreaming about, they found that those spikes of activity were associated with specific dream contents. Dreams emphasizing thinking tend to show brain activity in the frontal brain regions, while those involving perceptual experiences tend to involve regions towards the back of the brain – in much the same way as the same experiences would in waking life. Going further, they found that dreams involving faces show activity in the fusiform face area (see Chapter 6) and those involving speech show activity in and around Wernicke's area. So it appears that what we experience while we are dreaming really does mirror what we experience in waking life – at least as far as the brain is concerned.

If that's the case, then why don't we act out our dreams? It's because sleeping itself is mediated by an older part of the brain, the pons. When we are asleep, the pons releases neurotransmitters which inhibit motor neurones. Those are the nerve cells that carry messages from the brain to the muscles, so without them we can't move our limbs. This means that while we are asleep the body is in a sort of paralysis, while the brain continues to be active. When we wake up, different neurotransmitters activate our motor neurones, so we can move again. In some rare cases, people can wake up without this happening, so they experience a temporary paralysis, which can be quite alarming. The opposite is happening with sleepwalkers, who may act out their dreams, or part of them, physically, because their motor neurones are not fully inhibited.

Drugs and consciousness

Sleeping isn't the only type of change in consciousness that we experience. There are various types of drugs that change consciousness, and, as far as we can tell, all human societies use drugs to do this in some way or another. In Western society alcohol is the main socially acceptable consciousness-changing drug; some other societies use marijuana or coca leaves as socially acceptable ways of changing consciousness. And, of course, there are many other drugs, some of which derive from plant or fungal sources and have been used traditionally in special ceremonies and rituals, others of which have been developed by modern chemists, and most of which are illegal in modern Western societies.

Drugs don't affect a single area of the brain. Instead, they work by changing the balance of brain chemicals – neurotransmitters – as they pass messages from one nerve cell to another. As we saw in Chapter 2, each junction between nerve cells responds to a distinctive neurotransmitter, and groups of neurones responding to particular chemicals form neural pathways around the brain. Some of these, like the reward pathways we looked at in Chapter 8, are fairly well understood, while others, particularly those influenced by specific drugs, have been less clearly mapped out.

Drugs change the balance of neurotransmitters in the brain in several different ways. Some of them reduce the level of the neurotransmitter by 'keying in' to neural receptors, blocking the normal neurotransmitter so it is not picked up. Nicotine works in this way: it is picked up at acetylcholine receptor sites in the muscles, which would normally receive action messages from the brain. Blocking these receptors makes the person feel more sluggish and less inclined to move. Nicotine is usually taken by smoking, and the carbon monoxide involved reduces oxygen uptake in the lungs, which also reduces energy. These two effects combine to make the brain interpret the effects of the drug as similar to relaxation. It isn't the same as the neural state we associate with relaxation, but it does offer a slightly different state of consciousness.

Other changes in consciousness are achieved by drugs very similar to naturally occurring neurotransmitters. They don't block the neural activity; instead, they stimulate it in the way a natural neurotransmitter might. Marijuana and the opioid drugs (heroin, morphine and so on) work in this way. Marijuana, for example, is very similar in structure to the brain chemical anadamide and is picked up at the same receptors, influencing both pain and perceptual pathways in the brain. The drug extracted from the cannabis plant is now widely used in therapy for HIV, multiple sclerosis and the management of chronic pain, as well as acting to ameliorate the effects of chemotherapy for people undergoing cancer treatments. Its medical use is complicated by the fact that cannabis is illegal as a recreational drug. Short-term recreational users enjoy the sense of dissociation and perceptual distortion which it can provide, although there is some evidence that long-term use can result in depression, possibly linked with the inertia typical of recreational users.

The opioid drugs, like morphine and heroin, are also taken up at neurotransmitter sites and they also stimulate the relevant neurones in the same way that 'natural' neurotransmitters do. The natural neurotransmitters in this case are the endorphins and enkephalins that we produce in response to vigorous exercise, to trauma, or when we are experiencing the 'fight or flight' response (see Chapter 12). They reduce physical sensation, allowing the body to deal with potentially painful injuries or to overcome the pain produced by strenuous exercise. In doing so, they can produce a sense of euphoria, or 'floating', which is the reason why people often feel good after exercising. Opiate drugs mimic this response, producing similar sensations, but, being synthetic, they also have side effects and can easily result in addiction. Some have argued that exercise can be addictive, too, for the same reasons, and this may be true. But, as a rule, we would consider addiction to exercise to be beneficial to the person rather than harmful.

After a neurotransmitter has been released into a synapse, to pass a message from one neurone to another, it is re-absorbed into the nerve cell so its effects are neutralized. Some drugs prevent this reuptake, leaving the neurotransmitter in the synapse so that it continues to stimulate the next neurone. Cocaine is one of these drugs: it blocks the reuptake of noradrenaline and dopamine, so that there are more of those chemicals in the brain than usual. This can produce an immediate sensation of increased energy and self-confidence, but it also damps down the overall amount of brain activity. In the short term, cocaine is relatively benign and is commonly used as a local anaesthetic, although its use can still be dangerous if it is used to cover up physical ill health or exhaustion. Heavy use can produce hallucinations and delusions, and in the long term it can be an addictive drug. These effects are exaggerated by the derivative known as 'crack cocaine', which passes into the brain more quickly (but is more addictive for the same reason), and is known to produce high levels of paranoia and aggression. Withdrawal from cocaine addiction produces feelings of emotional pain as well as physical symptoms.

The prosocial drug MDMA, or 'ecstasy', also acts to block the reuptake of a neurotransmitter – in this case serotonin. Unlike crack cocaine, though, MDMA improves mood and empathy and has been shown to have a particularly strong effect on how we respond to socially based emotional stimuli. It's not the emotion itself:

studies of the way people change their responses under MDMA show that they don't react any more strongly to other kinds of emotional stimuli or to non-social content (Wardle, Kirkpatrick and de Wit, 2014). But it enhances our sense of closeness with others and our positive reactions to social contact. Before it was made illegal, this drug was sometimes used in marriage guidance counselling. It helped couples to be able to talk freely to one another without their negative emotions getting in the way.

Figure 13.5 MDMA enhances social closeness.

MDMA doesn't just block reuptake. It also stimulates the brain into releasing increased levels of particular neurotransmitters, including serotonin, involved in mood regulation; noradrenaline (norepinephrine), which stimulates alertness and moderate arousal; dopamine and oxytocin. Dopamine, as we've seen, is active in the reward pathways of the brain, and oxytocin, as we saw in Chapter 9, is particularly involved in affiliation and relationships. Putting all these effects together produces a powerful prosocial effect on our awareness.

Drugs, then, can act in many different ways to influence how we feel or how we see the world around us; and they produce many different effects on our consciousness. This raises the question: what exactly are we changing when we alter our states of consciousness? Like any attempt to define consciousness, it's difficult to pin down: there are many different aspects of consciousness. Any given drug

Functions involved in altered states of consciousness (adapted from Farthing, 1992)	
1. Attention	8. Time perception
2. Perception	9. Emotional feeling and expression
3. Imagery and fantasy	10. Arousal
4. Inner speech	11. Self-control
5. Memory	12. Suggestibility
6. Higher-level thought processes	13. Body image
7. Meaning and significance	14. Sense of personal identity

might affect only a couple of them, or it might affect many. The table below is a list of different aspects of consciousness, any or all of which might be influenced by drugs or by other life experiences like exercise or meditation.

Case study: Phineas Gage and the tamping iron

Probably the most famous case in the whole of neuroscience is that of Phineas Gage. Working as a railroad foreman in 1848, Gage suffered an explosives accident in which a tamping iron was driven right through his brain. It entered through the roof of his mouth and came out through the top of his skull, passing right through the orbitofrontal cortex and landing 25 yards away. There were many striking things about this accident, not least of which was that it didn't kill him. He didn't even lose consciousness: he was talking within a couple of minutes of the accident, sat upright while the horse and buggy drove the mile to his lodging, and was able to talk with the doctors who attended him. He recovered, but friends and relatives insisted that his personality had changed. He was more impatient, capricious and restless, and inclined to swear easily. This was attributed to the damage to his frontal lobes; but the fact that he retained his consciousness and sense of self throughout was remarkable.

Social consciousness and humour

It's no surprise that the functions involved in altered states of consciousness shown in the table above include higher-level thought processes and emotional expression. They, too, correlate with activity in particular parts of the brain. We have seen how the brain helps us to interact with other people, with special areas for recognizing faces, language and even body posture. We've also seen how the frontal lobes of the brain are active in the cognitive aspects of day-to-day living and in decision-making, and how they process many of our social interactions with other people.

This becomes very apparent if we look at one of the aspects that we find particularly special in our interactions – humour. We use humour in a variety of ways: to entertain, to express social familiarity, to lighten a tense situation, to re-examine what is going on, and even to help us to cope with traumatic events. By allowing a different, but emotionally safe, 'take' on a situation, humour allows us to reappraise situations and to generate or re-establish positive interactions with other people.

Psychologists distinguish between two aspects of humour: the cognitive aspect, which has to do with understanding; and the affective aspect, which has to do with the feelings of amusement and enjoyment resulting from that understanding. Having an active sense of humour is therapeutic: it has been shown to increase the effectiveness of our immune systems, and also to benefit the activity of the central nervous system as a whole. It also helps to reduce sensations of pain and discomfort, probably through the distraction caused by paying attention to the funny situation.

It is generally believed that humour is processed in the right side of the brain, but in fact both hemispheres are involved in the process. Humour as a whole involves the prefrontal cortex – part of the frontal lobes – but different aspects of it involve other areas of the brain as well. For example, detecting and appreciating humour mainly involves the temporal and prefrontal lobes of the brain. Part of this is semantic analysis - detecting the meanings and implications of words or actions. These, as we saw in Chapter 10, are processed in the inferior parietal lobule on the left temporal lobe, which contains Wernicke's area, where we decode speech. This area is close to and connects with the temporal gyrus, which monitors rules of language. It detects contradictions, things out of context, or broken rules. Both of these areas become active when we detect verbal information that is incongruous or inappropriate – a feature of many humorous situations.

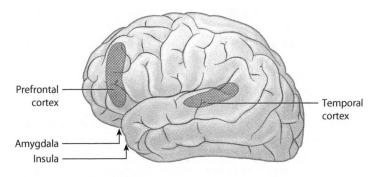

Prefrontal cortex

Temporal cortex

Amygdala

Insula

Figure 13.6 Humour in the brain

Non-verbal humour, such as that in cartoons and slapstick, involves slightly different brain areas. Samson, Zysset and Huber (2008) used fMRI to study the brain areas involved in humorous cartoons, and found that the incongruity is still processed primarily in the left hemisphere, but this time in the area that crosses the temporal and parietal lobes and in the prefrontal cortex. Appreciating humour itself involves the amygdala and the insula, in both hemispheres. We've seen how both of these areas are active in emotions, and so they set off the affective element in humour – the bit that is all about smiling and enjoyment. And, as we know from our own experience, that can make all the difference to how information is perceived.

Being human, then, is much more than simply being a processor of information. Computers can do that. Our marvellous brains allow us to deal with vast amounts of information, shaping and directing that information so that we can use it effectively without becoming overwhelmed, letting us plan, think and decide, and allowing us to be aware of ourselves. And, unlike computers, they do all that within the context of emotional nuance and social awareness. We rightly recognize the cybermen of *Dr Who*, with all their emotions 'deleted', as monsters. Being human is to experience warmth, empathy, kindness and many other emotional nuances, negative as well as positive. It is the way that our brains regulate our emotions, and gather and use our social knowledge, which makes this possible.

Focus points

✳ Human decision-making, unlike computer decision-making, is strongly influenced by bias. This reflects our evolutionary history and survival needs, and also our social natures.

✳ Different types of consciousness are reflected in different patterns of brain activity. There is no single area for consciousness, but the claustrum is important and has many connections with both cortical and subcortical brain structures.

✳ EEG measures show different levels of sleep. Dream activity is reflected by activity in the relevant parts of the cortex, but inhibitory synapses in the pons mean that they do not affect the body.

✳ Psychoactive drugs affect consciousness by either changing the activity of neurotransmitters or imitating them.

✳ Humour is processed on both sides of the brain and it activates a wide area of the cerebral cortex. It also involves the positive emotion areas of the amygdala and the insula.

Next step

This book has described some of the ways that the brain works, but what we have been able to cover here is only the tip of the iceberg. Every day, researchers are discovering something new about brain functioning, and each technical advance tells us even more. It's a fascinating area, and I hope you have enjoyed this overview of some of the main discoveries and that you will find out more as more discoveries are made.

Glossary

acalculia Disorder caused by brain damage in which people are unable to understand or manipulate number. Compare: dyscalculia.

achromatopsia Rare disorder of the human brain that causes people to see the world only in shades of grey.

agraphia Disorder in which people have difficulty with writing.

alpha rhythms Distinct, low-amplitude, low-frequency rhythmical patterns of electrical activity that occur in the brain during periods of relaxation. Also referred to as alpha waves.

amusia Inability to process music.

amygdala (Gk. 'almond') The two almond-shaped structures in the brain that are actively involved in processing emotions, notably fear.

angular gyrus Part of the fusiform face area in the left hemisphere that is involved in reading, especially with the identification of letters and words. Compare: fusiform gyrus.

anomic aphasia Difficulty in finding the right words.

anosmia Inability to perceive smells.

anterior Term used to describe structures near the front of the brain. Compare: posterior.

anterior cingulate cortex Area of the brain that evaluates risks and works out whether an action is likely to be rewarded or punished.

anterior paracingulate cortex Area of the brain that is involved in thinking about other people's intentions.

aphasia Problems with language.

arcuate fasciculus Bundle of nerve fibres connecting Broca's and Wernicke's areas.

axon Long projection of a neuron conducting the electrical impulse away from the nerve cell body and towards another neuron.

basal ganglia Group of cells in the white matter of the frontal lobes which help to organize and inhibit movement, and are active in reward pathways.

binocular disparity Difference between the images produced by the two eyes that helps the brain to process distance.

biological motion Movement produced by living bodies.

blindsight Phenomenon by which a blind person nonetheless responds to visual stimuli.

Broca's area Area of the brain that helps with the formation of speech.

Capgras syndrome Condition, possibly caused by brain damage, under which an individual becomes convinced family members have been replaced by strangers.

CAT or CT (computerized tomography) scan Brain scan that uses a series of X-ray or ultrasound images of the brain to form a 3D image.

cerebellum Sometimes known as the 'mini-brain', the distinctive wrinkled bulge under the back of the cerebrum that coordinates physical movement.

cerebral cortex Outer layer of the cerebrum.

cerebral hemispheres The two halves of the cerebrum, bound together by the corpus callosum.

cerebrum The largest structure of the mammalian brain, associated with thinking, perceiving, language, imagining, planning, decision making, consciousness, etc. The associated adjective is **cerebral**.

cingulate cortex Large area of the brain above the corpus callosum involved in emotions, memory and learning; often considered part of the limbic system.

claustrum Thin layer of cells that links many areas of the cerebrum and connects to the limbic system; in some experts' opinion, a structure associated with human consciousness.

colour constancy The ability to perceive colours as constant even under varying light conditions.

conduction aphasia Inability to repeat spoken language, or to read aloud accurately.

connector neurones See: interneurones.

corpus callosum The flat bundle of fibres that connects and enables communication between the two hemispheres of the brain.

cortex The outer layer of various parts of the brain. The associated adjective is **cortical**.

dementia Neurocognitive disorder that results in a gradual impairment of a person's ability to think and remember.

diencephalon Group of subcortical structures including the thalamus, hypothalamus, the pineal gland and the pituitary gland.

dorsal Towards the top.

dorsal stream Neural pathway connecting auditory processing areas in the cerebral cortex that helps locate where sounds are coming from. Compare: ventral stream.

dorsal visual stream Neural pathway connecting the visual cortex with areas in the parietal lobe that helps locate objects. Compare: ventral visual stream.

dorsolateral prefrontal cortex (PFC) Area of the prefrontal cortex concerned with working memory.

dyscalculia Developmental disorder in which people are unable to understand or manipulate number. Compare: acalculia.

dyslexia Disorder in which people incorrectly recognize words.

efMRI See: magnetic resonance imaging.

electrical impulse Sudden burst of electricity in the brain that moves from neuron to neuron.

enactive representation Memory of actions or physical sensations, often known as 'muscle memory'.

entorhinal cortex The area around the hippocampus involved in transfering memories for long term storage.

equilibrioception Sense of balance.

event-related potential (ERP) Distinctive pattern in EEG measures of brain activity.

extrapyramidal motor system System in the brain which processes unconscious, automatic movement. Compare: pyramidal motor system.

extrastriate body area (EBA) Area of the brain just outside the visual cortex that responds to other people's body parts and outlines. Compare: occipital face area.

fight or flight response Physiological state that provides the body with a temporary boost of energy in the face of an external threat.

fissure Deep, narrow grove in the brain.

fMRI See: magnetic resonance imaging.

fusiform body area (FBA) Area of the brain underneath the visual cortex that seems to responds to the sight of human bodies, including their shape and size.

fusiform face area (FFA) Area of the brain underneath the visual cortex that responds to other people's faces, especially familiar ones. Compare: occipital face area.

fusiform gyrus A part of the fusiform face area in the left hemisphere that is involved in reading, especially with the interpretation of letters and words.

glial cells Brain cells that support the neurones by holding them in place and providing them with oxygen and nutrients.

globus pallidus Part of the basal ganglia involved in the regulation of voluntary movement

gustation Sense of taste.

gyrus (pl. gyri) Name for the mounds between the grooves in the cerebral hemispheres.

hippocampus (Gk. 'seahorse') Also known as the medial temporal lobe, the main centre for the consolidation and storage of memories in the brain.

homeostasis Maintenance of steady, comfortable conditions in the body.

hormone Chemical produced by a gland that helps regulate physiology and behaviour.

hypothalamus The small 'lump' just below (Gk. *hypo*) the thalamus that helps regulate body temperature as well as homeostasis.

inferior Towards the bottom, or underneath. Compare: superior.

inferior occipital gyrus Area in the optical cortex that helps with facial recognition.

inferior parietal lobule A large part of the brain where the occipital and parietal lobes meet, which contains areas known to process language.

instrumental aggression Aggression initiated by the person in order to achieve a particular goal.

insula An area of cortex tucked underneath the frontal lobes, associated with emotion and self-awareness.

interneurone Sometimes known as a connector neurone, a simple structure consisting of a cell body with extending branches (dendrites) that make connections between nerve cells

interoception Perception of movement and pain within the body.

kinaesthesia Sense of movement.

Korsakoff's syndrome Form of amnesia caused by damage to the hippocampus as a result of excessive alcohol consumption.

lateral Towards the side.

lateral premotor cortex Area at the side of the premotor cortex that prepares physical actions.

limbic system A collective name for a set of brain structures located around the thalamus, often active in emotional and behavioural responses.

lobe In the brain, one of four general sections of each cerebral hemisphere, or one of the two halves of the parts of the cerebellum. Generally used to refer to a rounded or fleshy lump.

magnetic resonance imaging (MRI) Brain-scanning technique that uses the magnetic fields produced by water molecules in brain cells to produce images of the brain. fMRI (functional magnetic resonance imaging) is used to explore specific brain activities/functions, while efMRI (event-related functional magnetic resonance imaging) compares the electrical activity produced by two or more events.

mechanoreception The sense of external pressure on the skin.

medial Towards or in the middle.

medial prefrontal cortex The middle part of the prefrontal cortex that contributes to the sense of self and identity.

medial temporal cortex The middle part of the temporal cortex, concerned with storing memories.

medulla Part of the brain at the top of the spinal cord, which regulates the body's basic, autonomic functions.

MEG scan (magneto-encephalography) Brain scan using SQUID (superconducting quantum interference devices) to detect changes in magnetic activity of the brain.

microelectrode Tiny electrode used that is able to record the activity of a single neuron.

midbrain Part of the brain above the brainstem which includes the pons and the reticular formation, and regulates alertness and some sensory reception.

mirror neuron A neuron that becomes active both when a person (or animal) performs an activity, and also when they observe the same action performed by another.

motor cortex Part of the cerebral cortex involved in the planning and coordinating of voluntary movement.

myelin sheath Fatty coating that covers an axon.

nucleus accumbens Area at the back of the amygdala associated with reward and positive reinforcement.

neural pathways Groups of neurons which carry electrical impulses around the brain along specific routes.

neuro- Combining prefix meaning 'nerve' or 'nervous system' (e.g. neuropsychology, neurotransmitter). The related adjective is **neural**.

neuroimaging Scanning techniques producing pictures or images of the brain.

neurone (or neuron) Nerve cell, carrying electrical signals.

neurotransmitter Chemical in the synaptic knob that provides the connection between neurones.

neural tube Simple tube that forms the core of the earliest nervous systems in animals.

nociception Perception of pain through the skin.

nociceptor Pain receptor that responds to mechanical, thermal or chemical stimuli.

occipital face area Part of the visual cortex that that responds to other people's faces. Compare: extrastriate body area.

olfactory bulb Parts of the brain used to interpret smells.

olfactory tubercle 'Control centre' in the olfactory cortex that receives smell-related information from many other areas of the brain.

optic chiasma Crossover point of the two optic nerves as they pass from the eye to the brain.

optical flow Changes in the visual images received by the brain, caused by the perceiver moving around in the physical world.

orbito- Close to, or at the level of, the eye sockets ('orbits').

orbitofrontal cortex Part of the frontal lobes immediately above the eye sockets, involved in attachments, motivation, and regulating social behaviour. Also known as the orbitofrontal gyrus.

paracingulate cortex Area of the cerebrum above the cingulate cortex, active in decoding and predicting social intentions.

periacqueductal gray Part of the midbrain particularly active in aggression.

PET (positron emission tomograph) **scan** Brain scan which traces the distribution of a radioactive chemical in the blood supply, taken up by active cells in the brain.

perirhinal cortex An area close to the hippocampus which is concerned with recognition and familiarity.

phantom pain Pain experienced in a part of the body that has been amputated.

phantosmia Experience of a non-existent smell.

photoreceptor Light-detecting cell in the retina.

phrenology Discredited practice of reading bumps on the head as a way of determining individual character and intelligence.

pinna Dish-like part of the outer ear that helps direct sound into the ear canal.

posterior Towards the back. Compare: anterior.

posterior paracingulate cortex The part of the paracingulate cortex which becomes active when thinking about personal interactions with others.

prefrontal cortex Area of the frontal lobe just behind the orbitofrontal cortex, concerned with choices and general intentions.

premotor cortex Part of the brain that prepares the motor cortex for action.

primary language pathway Neural pathway from the language areas processing understanding at the back of the temporal lobe to the motor areas in the frontal lobe involved in speech production.

proprioception Information received from muscles and joints.

prosopagnosia Inability to recognize faces.

pure alexia Disorder in which people can spell out words letter by letter but are unable to recognize the whole word.

pyramidal motor system System in the brain which processes deliberate movement. Compare: extrapyramidal motor system.

Raphé nuclei Cluster of nuclei in the brainstem active in reward pathways in the brain.

reactive aggression Aggression which arises from feelings of threat or frustration.

reflex Rapid muscle movement that occurs in response to a painful stimulus.

REM sleep Sometimes known as paradoxical sleep, a type of sleep that occurs during the first stage of the sleep cycle, characterized by rapid eye movements and dreaming.

Schwann cells Fatty cells which wrap around cell axons forming an insulating myelin sheath.

somatosensation The sense of touch, involving mechanoreception, thermoreception and nociception.

spinal cord Tube of nerve fibres running the length of the spine, linking the body's nerve fibres with the brain.

subcortex A general name for the parts of the brain below the cerebral cortex – that is, everything except the cerebrum itself. The related adjective is **subcortical**.

substantia nigra Part of the basal ganglia playing an important role in attachment and reward.

sulcus (pl. sulci) A groove, or fissure, in the cerebral cortex.

superior Towards the top, or above. Compare: inferior.

superior temporal sulcus (STS) Groove at the top of the temporal lobes defining an area concerned with facial recognition and detecting and interpreting social cues.

supplementary motor area (SMA) Part of the premotor cortex that receives proprioceptive information about how parts of the body are positioned.

supramarginal gyrus Part of the fusiform face area concerned with word choices and also empathy.

synapse The connecting gap between two neurones. The related adjective is **synaptic**.

synaptic plasticity The ability of synapses to grow and become more effective when frequently activated.

system 1 thinking Rapid, intuitive thinking, often inaccurate and subject to bias.

system 2 thinking Logical, deliberate thinking, usually precise but slower.

thalamus Large structure below the cerebrum which relays motor and sensory signals to the cerebral cortex.

thermoreception Perception of temperature on the skin.

theta rhythm Distinct, high-amplitude rhythmical patterns of electrical activity in the brain during normal waking activity.

tip-of-the-tongue phenomenon (TOT) The common experience of knowing we know something but not being quite able to remember it.

tonotopic map Conceptual map of high- and low-pitched sounds produced by the primary auditory cortex.

transcortical aphasia Group of aphasias involving different areas across the cortex and interfering with speech production and recognition.

transcranial magnetic stimulation (TMS) Technique of studying the brain using short bursts of magnetic stimulation, which can influence very localized areas of the cortex.

transcranial direct current stimulation (tDCS) Technique of studying the brain using an electrical coil held directly over the scalp, producing a 'virtual lesion' interfering with brain function.

transduction The conversion of sensations into electrical impulses that can be passed on to the brain.

trapezoid body The area in the pons where the auditory nerves from the two ears cross over.

ventral Towards the bottom.

ventral stream Neural pathway connecting auditory processing areas in the cerebral cortex which helps identify what sounds represent. Compare: dorsal stream.

ventral striatum Part of the basal ganglia near to the cerebellum involved in attachments, positive emotions and rewards.

ventral visual stream Neural pathway connecting the visual cortex with areas in the temporal lobe that helps identify what objects are. Compare: dorsal visual stream.

ventrolateral prefrontal cortex The part of the prefrontal cortex at the side and towards the bottom, involved in cognitive control and rule following.

ventromedial prefrontal cortex The part of the prefrontal cortex towards the bottom and tucked into the middle of the brain, involved in processing fear and risk and regulating aggressive behaviour.

vestibular system Part of the inner ear that detects spatial orientation and movement.

ventral tegmental area (VTA) Group of neurons in the midbrain that play an important part in the dopamine reward pathway in the brain.

visual field The total area of the image received by the eyes when they are focused on a single point.

Wernicke's area Area of the brain involved in the understanding of others' speech. Compare: Broca's area.

Index

Image credits

1.1 The human brain © Shutterstock.com

2.5 An MRI scanner © Shutterstock.com

2.6 A typical brain scan © Shutterstock.com

4.6 Dance happens in all cultures © Shutterstock.com

6.6 Musical performance © Shutterstock.com

8.1 Emotional facial expressions © Shutterstock.com

13.5 Prosocial activity © Shutterstock.com

Also by Dr Nicky Hayes

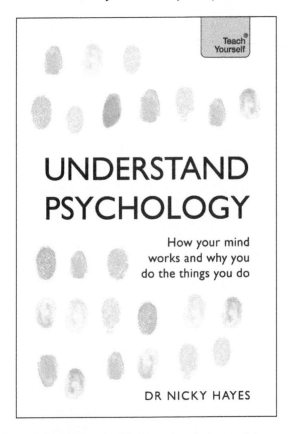

A fascinating insight into what makes us tick
The bestselling *Understand Psychology* explains basic psychological processes and how they influence us in all aspects of everyday life. It explores why we are the way we are, how we came to be that way, and what we might do to change seemingly fundamental traits.

The book puts psychology in context, using non-technical language to analyse everyday situations. It is a comprehensive introduction that shows how human experience can be understood on many levels.

ISBN: 9781444100907